INCREDIBLE ADVENTURE AND EXPLORATION STORIES

TALES OF DARING FROM ACROSS THE GLOBE

EDITED BY VERONICA ALVARADO

Skyhorse Publ...

D1444998

Skyhorse Publishing books may be purchased in bulk at special discounts for sales promotion, corporate gifts, fund-raising, or educational purposes. Special editions can also be created to specifications. For details, contact the Special Sales Department, Skyhorse Publishing, 307 West 36th Street, 11th Floor, New York, NY 10018 or info@skyhorsepublishing.com.

Skyhorse® and Skyhorse Publishing® are registered trademarks of Skyhorse Publishing, Inc.®, a Delaware corporation.

Visit our website at www.skyhorsepublishing.com.

10 9 8 7 6 5 4 3 2 1

Library of Congress Cataloging-in-Publication Data is available on file.

Cover design by Tom Lau
Cover photo credit: iStockphoto

Print ISBN: 978-1-5107-3223-0
Ebook ISBN: 978-1-5107-3228-5

Printed in China

TABLE OF CONTENTS

INTRODUCTION

I'd argue that you'd be hard-pressed to find a recent film that is more crowd-pleasing, uplifting, or family-friendly than Disney-Pixar's 2009 animated feature *Up!* The movie follows crotchety old widower Carl Frederickson, who since the death of his beloved wife Ellie, has forsaken his once adventurous youthful spirit and has stubbornly chosen to stick to his boring but safe routine. But Carl is shaken out of complacency when he's visited by precocious, eight-year-old "Wilderness Explorer" Russell. Through a series of zany circumstances that takes the pair to exotic lands and on heart-pounding adventures, Russell reminds Carl that life can't be lived just sitting on your couch. As he tells Carl (and thus the viewer), "Adventure is out there!"

Now, most of our lives are spent going about our daily routine: waking up, heading to work or school, and closing our eyes at the end of the day, only to repeat it all again come the next morning. There is rarely any variation; which is why we often find ourselves longing for a break from the monotony. We yearn to get of town, hop on the nearest plane, train, or automobile, and seek escape in an unfamiliar territory. For most of us, these journeys can only take place in our dreams or on sporadic vacations. But there are some lucky men and women throughout history whose mottoes echo Russell's. Whether trekking to the world's highest peaks, touching down on its most remote corners, or journeying to the sea's darkest depths, these explorers, throughout time, have stared into the unknown on a regular basis.

This collection celebrates the spirit of adventure that inhabits all of us. Divided into three sections, it features both true and fictional tales of adventure and exploration that span the world. The first section contains four legendary journeys taken across the sea. From Viking conquests to the present day, mankind has embraced its nautical spirit, and sought to discover what lies on the other side of the horizon. But as the second section shows, journeys across the land can prove to be just as adventure-filled, be it in the form of dangerous treks across the desert, death-defying climbs over mountain peaks, or terrifying encounters with beasts. The final section contains celebrated fictional adventures from four of Western literature's classic writers.

So while these adventure stories may alternately delight, terrify, or astound you, I hope that they inspire you to undertake some quests of your own, both near and far. After all, "adventure is out there!"

—Veronica Alvarado
Spring 2018

PART ONE

ADVENTURES BY SEA

CHAPTER 1

DRAKE'S CIRCUMNAVIGATION

by Richard Hakluyt

The Golden Hinde. A full-scale replica of Drake's original vessel, the *Golden Hind*, docked at St. Mary Ovary Dock, London.

Editor's Note: Please note that for the sake of preserving the authenticity of the text, written in 1641, Hakluyt's original spelling, grammatical, and stylistic choices have been preserved.

The famous voyage of Sir Francis Drake into the South sea, and therehence about the whole Globe of the earth, begun in the yeere of our Lord 1577.

The 15 day of November, in the yeere of our Lord 1577, M. Francis Drake, with a fleete of five ships and barkers, and to the number of 164 men, gentlemen and sailers, departed from Plimmouth, giving out his pretended voyage for Alexandria: but the wind falling contrary, hee was forced the next morning to put into Falmouth haven in Cornewall, where such and so terrible a tempest tooke us, as few men have seene the like, and was in deed so vehement, that all our ships were like to have gone to wracke: but it pleased God to preserve us from that extermitie, and to afflict us onely for that present with these two particulars: The mast of our Admirall which was the Pellican, was cut over boord for the safegard of the ship, and the Marigold was driven ashore, and somewhat bruised for the repairing of which damages we returned agains in Plimmouth, and having recovered those harmes, and brought the ships againe to good state, we set forth the second time from Plimmouth, and set saile the 13 day of December following.

The 25 day of the same moneth we fell with the Cape Cantin, upon the coast of Barbarie, and coasting along, the 27 day we found an Island called Mogador, lying one mile distant from the maine, betweene which Island and the maine, we found a very good and safe harbour for our ships to the in, as also very good entrance, and voyde of any danger.

On this Island our Generall erected a pinnesse, whereof he brought out of England with him foure already framed, while these things were in doing, there came to the waters side some of the inhabitants of the countrey, shewing foorth their flags of truce, which being seene of our Generall, hee sent his ships boate to the shore, to know what they would: they being willing to come aboord, our men left there one man of our company for a pledge, and brought two of theirs aboord our ship, which by signes shewed our Generall, that the next day they would bring some provision, as sheepe, capons and hennes, and such like: whereupon our Generall bestowed amongst them some linnen cloth and shooes, and a javeling, which they very joyfully received, and departed for that time.

The next morning they failed not to come againe to the waters side, and our Generall againe setting out our boate, one of our men leaping over rashly ashore, and offering friendly to imbrace them, they set violent hands on him, offering a dagger to his throte if hee had made any resistance, and so laying him on a horse, caried him away: so that a man cannot be too circumspect and warie of himselfe among such miscreants.

Our pinnesse being finished, wee departed from this place the 30 and last day of December, and coasting along the shore, we did descrie, not contrary to our expectation, certaine Canters which were Spanish fishermen, to whom we gave chase and tooke three of them, and proceeding further we met with 3 caravels and tooke them also.

The 17 day of January we arrived at Cape Blanco, where we found a ship riding at anchor, within the Cape, and but two simple Mariners in her, which ship we tooke and caried her further into the harbour, where we remained 4 dayes, and in that space our Generall mustered, and trayned his men on land in warlike maner, to make them fit for all occasions.

In this place we tooke of the Fishermen such necessaries as we wanted, and they could yeeld us, and leaving heere one of our litle barkes called the Benedict, wee tooke with us one of theirs which they called Canters, being of the burden of 40 tunnes or thereabouts.

All these things being finished, wee departed this harbour the 22 of Januarie, carying along with us one of the Portugall Caravels which was bound to the Island of Cape Verde for salt, whereof good store is made in one of those Islands.

The master or Pilot of that Caravel did advertise our Generall that upon one of those Islands called Mayo, there was great store of dryed Cabritos, which a few inhabitants there dwelling did yeerely make ready for such of the Kings Ships as did there touch being bound for his countrey of Brasile or elsewhere. We fell with this Island the 27 of January, but the Inhabitants would in no case traffique with us, being thereof forbidden by the Kings Edict: yet the next day our Generall sent to view the Island, and the likelihoods that might be there of provision of victuals, about threescore and two men under the conduct and government of Master Winter and Master Doughtiue, and marching towards the chiefe place of habitation in this Island (as by the Portugall wee were informed) having travailed to the mountains the space of three miles, and arriving there somewhat before the day breake, we arrested our selves to see day before us, which appearing, we found the inhabitants to be fled: but the place, by reason that it was manured, we found to be more fruitfull then the other part, especially the valleys among the hils.

Here we gave our selves a little refreshing, as by very ripe and sweete grapes, which the fruitfulnesse of the earth at that season of the yeere yielded us: and that season being with us the depth of Winter, it may seeme strange that those fruites were then there growing: but the reason thereof is this, because they being betweene the Tropike and the Equinoctiall, the Sunne passeth twise in the yeere through their Zenith over their heads, by meanes whereof they have two Summers, & being so neere the heate of the line they never lose the heate of the Sunne so much, but the fruites have their increase and continuance in the midst of Winter. The Island is wonderfully stored with goates and wilde hennes, and it hath salt also without labour, save onely that the people gather it into heapes, which continually in great quantitie is increased upon the sands by the flowing of the sea, and the receiving heate of the suns kerning the same, so that of the increase thereof they keeps a continuall traffique with their neighbours.

Amongst other things we found here a kind of fruit called Cocos, which because it is not commonly knowen with us to England, I thought good to make some description of it.

The tree beareth no leaves nor branches, but at the very top the fruit growth in clusters, hard at the top of the branch of the tree, as big every severall fruite as a mans head: but kerning taken off the uttermost barke, which you shall find to bee very full of strings or sinowes, as I may terme them, you shall come to a hard shell which may holde of quantitie in liquor a pint commonly, or some a quart, and some lesse: within that shell of the thickness of halfe an inch good, you shall have a kinde of hard substance and very white, no lesse good and sweete then almonds: within that againe a certaine cleare liquor, which being drunke, you shall not onely finde it very delicate and sweete, but most comfortable and cordiall.

After we had satisfied our selves with some of these fruites, wee marched further into the Island, and saw great store of Cabritos alive, which were so chased by the inhabitants, that we could doe no good towards our provision, but they had layde out as it were to stoppe our mouths withal, certaine olde dryed Cabritos, which being but ill, and small and few, we made no account of it.

Being returned to our ships, our Generall departed hence the 31 of this moneth, and sayled by the Island of S. Iago, but farre enough from the danger of the inhabitants, who shot and discharged at us three peeces, but thy all fell short of us, and did us no harme. The Island is fayre and large, and as it seemeth, rich and fruitfull, and inhabited by the Portugals, but the mountains and high places of the Island are sayd to be possessed by the Moores, who having bin slaves to the Portugals, to ease themselves, made escape to the desert places of the Island, where they abide with great strength.

Being before this Island, we espied two ships under sayle, to the one of which we gave chase, and in the end boorded her with a ship-boat without resistance, which we found to be a good prize, and she yielded unto us good store of wine: which prize our Generall committed to the custodie of Master Doughtie, and retaining the Pilot, sent the rest away with his pinnesse, giving them a Butte of wine and some victuals, and their wearing clothes, and so they departed.

The same night we came with the Island called by the Portugals, IIha del fogo, that is, the burning Island: in the Northside whereof is a consuming fire, the matter is sayde to be of Sulphure, but notwith-standing it is like to be a commodious Island, because the Portugals have built, and doe inhabite there.

Upon the south side thereof lyeth a most pleasant and sweete Island, the trees whereof are always greene and faire to looke upon, in respect whereof they cakk it Ilha Brava, that is, the brave Island. From the bankes thereof into the sea doe run in many places reasonable streames of fresh waters easie to be come by, but there was no con-venient roade for our ships: for such was the depth, that no ground could be had for anchoring, and it is reported, that ground was never found in that place, so that the tops of Fogo burne not so high in the ayre, but the rootes of Brava are quenched as low in the sea.

Being departed from these Islands, we drew towards the line, where we were becalmed the space of 3 weekes, but yet subject to divers great stormes, terrible lightnings and much thunder: but with this miserie we had the commoditie of great store of fish, as Dolphins, Bonitos, and flying fishes, whereof some fell into our shippes, where-hence they could not rise againe for want of moisture, for when their wings are drie, they cannot flie.

From the first day of our departure from the Islands of Cape Verde, wee sayled 54 days without sight of land, and the first land that we fell with was the coast of Brasil, which we saw the fift of April in ye height at 33 degrees towards the pole Antarctike, and being discovered at sea by the inhabitants of the countrey, they made upon the coast great form for a sacrifice (as we learned) to the devils, about which they use conjurations, making heapes of sande and other ceremonies, that when any ship shall goe about to stay upon their coast, not onely sands may be gathered together in shoalds in every place but also the stormes and tempests may arise, to the casting away of ships and men, whereof (as it is reported) there have bene divers experiments.

The seventh day in a mightie great storme both of lightning, rayne and thunder, we lost the Canter which we called the Christopher: but the eleventh day after, by our Generalls great care inn dispersing his ships, we found her againe, and the place where we met, our Generall called the Cape of Joy, where every ship tooke in some water. Here we found a good temperature and sweet ayre, a very faire and pleasant countrey with an exceeding fruitfull soyle, where were great store of large and mightie Deere, but we came not to the sight of any people: but traveiling further into the countrey, we perceived the footing of the people in the clay-ground, shewing that they were men of great stature. Being returned to our ships, we wayed anchor, and ranne somewhat further, and harboured our selves betweene a rocke and the maine, where by meanes of the rocke that brake the force of the sea, we rid very safe, and upon this rocke we killed for our provision certaine sea-wolves, commonly called with us Seales.

From hence we went our course to 36 degrees, and entred the great river of Plate, and ranne into 54 and 55 fadomes, and a halfe of fresh water, where wee filled our water by the ships side: but our Generall finding here no good harborough, as he thought he should, bare out againe to sea the 27 of April, and in bearing out we lost sight of our Flieboate wherein Master Doughtie was, but we saying along, found a fayre and reasonable good Bay wherein were many, and the same profitable Islands, one whereof had so many Seales, as would at the least have laden all our shippes, and the rest of the islands are as it were laden with foules which is wonderfull to see, and they of divers sortes. It is a place very plentifull of victuals, and hath in it no wants of fresh water.

Our Generall after certaine dayes of his abode in this place, being on shore in an Island, the people of the countrey shewed themselves unto him, leaping and dauncing, and entred into traffique with him, but they would not received anything at any mans hands, but the same must bee cast upon the ground. They are so cleane, comely, and strong bodies, swift on foote, and seeme to be very active.

The eighteenth day of May our Generall thought it needful to have a care of such Ships as were absent and therefore indeavouring to seeke the Fleinboate wherein Master Doughtie was, we espied here againe the next day: and whereas certaine of our ships were sent to discovere the coast and to search an harbour, the Marygold and the Canter being imployed in that businesse, came unto us and gave us understanding of a safe harbour that they had found, wherewith all our ships bare, and entred it, where we watered and made new provisions of victuals, as by Seales, whereof we slew to the number of 200 or 300 in the space of an houre.

Here our Generall in the Admiral rid close aboord the Flieboate, and tooke out of her all the provisions of victuals and what els was in her, and hailing her to the Lande, set fire to her, and so burnt her to save the iron worke: which being a doing, there came downe of the countrey certaine of the people naked, saving only about their waste the skinne of some beast with the furre or haire on, and something also wreathed on their heads: their faces were painted with divers colours, and some of them had on their heads the similitude of hornes every man his bow which was an ell in length, and a couple of arrows. They were very agill people and quicke to deliver, and seemed not to be ignorant in the feates of warres, as by their order of ranging a few men, might appeare. These people would not of a long time receive any thing at our handes; yet at length our Generall being ashore, and they dancing after their accustomed maner about him, and hee once turning his backe towards them, one leapt suddenly to him, and tooke his cap with his golde band off his head, and ran a little distance from him and shared it with his fellow, the cap to the one, and the band to the other.

Having dispatched all our businesse in this place, wee departed and set sayle, and immediately upon our setting foorth we lost our Canter which was absent three or foure dayes but when our Generall had her againe, he tooke out the necessaries, and so gave her over neere to the Cape of Good hope.

The next day after being the twentieth of June, wee harboured our selves againe in a very good harborough, called by Magellan Port S. Julian, where we found a gibbet standing upon the maine, which we supposed to be the place where Magellan did execution upon some of his disobedient and rebellious company.

The two and twentieth day our Generall went ashore to the maine, and in his companie, John Thomas, and Robert Winterhie, Oliver the Master gunner, John Brewer, Thomas Hood, and Thomas Drake, and entering on land, they presently met with the two or three of the countrey people, and Robert Winterhie having in his hands a bowe and arrows, went about to make a shoote of pleasure, and in his draught his bowstring brake, which the rude Savages taking as a token of warre, began to bend the force of their bowes against our company, and drove them to their shifts very narrowly.

In this Port our Generall began to enquire diligently of the actions of M. Thomas Doughtie, and found them not to be such as he looked for, but tending rather to contention or mutinie, or some other disorder, whereby (without redresse) the successe of the voyage might greatly have been hazarded: whereupon the company was called together and made acquainted with the particulars of the cause, which were found partly by Master Doughties owne confession, and partly by the evidence of the fact, to be true: which when our Generall saw, although his private affection to M. Doughtie (as hee then in the presence of us all sacredly protested) was great, yet the care he had of the state of the voyage, of the expectation of her Majestie, and of the honour of his countrey did more touch him, (as indeede it ought) then the private respect of one man: so that the cause being thoroughly heard, and all things done in good order as neere as might be the course of our lawes in England, it was concluded that M. Doughtie should receive punishment according to the qualitie of the offence: and he seeing no remedie but patience for himselfe, desired before his death to receive the Communion, which he did at the hands of M. Fletcher our Minister, and our Generall himselfe accompanied him in that holy action: which being done, and the place of execution made ready, hee having embraced our Generall and taken his leave of all the companie, with prayer for the Queenes majestie and our realme, in quiet sort laid his head to the blocke, where he ended his life. This being done, our Generall made divers speeches to the whole company, perswading us to unitie, obedience, love, and regard of our voyage; and for the better confirmation thereof, willed every man the next Sunday following to prepare himself to receive the Communion, as Christian brethren and friends ought to doe, which was done in very reverent sort, and so with good contentement every man went about his businesee.

The 17 day of August we departed the port of S. Julian, & the 20 day we fell with the straight or freat of Magellan going into the South sea, at the Cape or headland whereof we found the bodie of a dead man, whose flesh was cleane consumed.

The 21 day we entered the straight, which we found to have many turnings, and as it were shuttings up, as if there were no passage at all, by meanes whereof we had the wind often against us, so that some of the fleete recovering a Cape or point of land, others should be forced to turne backe againe, and come to an anchor where they could.

In this straight there be many faire harbors, with store of fresh water, but they lacke their best commoditie: for the water is there of such depth, that no man shal find ground to anchor in, except it bee in some narrow river or corner, or betweene some rocks, so that if any extreme blasts or contrary winds do come (whereunto the place is much subject) it careith with it no small danger.

The land on both sodes is very huge & mountainous, the lower mountains whereof, although they be monstrous and wonderfull to looke upon for their height, yet there are others which in height exceede them in a straight maner, reaching themselves a bvoe their fellows so high, that betweene them did appeare three regions of cloudes.

These mountains are covered with snow: at both the southerly and Easterly of the straight there are Islands, among which the sea hath his indraught into the streights, even as it bath in the maine entrance of the freat.

This straight is extreme cold, with frost and snow continually; the trees seeme to stoope with the burden of the weather, and yet are greene continually, and many good and sweete herbes doe very plentifully grow and increase under them.

The bredth of the straight is in some place a league, in some other places 2 leagues, and three leagues, and in some other 4 leagues, but the narrowest place hath a league over.

The 24 of August we arrived at an Island in the streights, where we found great store of foule which could not flie, of the bignesse of geese, whereof we killed in lesse then one day 3000 and victualled our selves thoroughly therewith.

The 6 day of September we entred the South sea at the Cape or head shore.

The seventh day wee were driven by a great storme from the entering into the South sea two hundred leagues and odde in longitude, and one degree to the Southward of the Streight: in which height, and so many leagues to the Westward, the fifteenth day of September fell out the Eclipse of the Moone at the houre of sixe of the clocke at night: but neither did the Eclipticall conflict of the Moone impayre our state, nor her clearing againe amend us a whit, but the accoustomed Eclipse of the Sea continued in his force, wee being darkened more then the Moone seven fold.

From the Bay (which we called The Bay of severing of friends) wee were driven backe to the Southward of the streights in 57 degrees and a terce: in which height we came to an anker among the Islands, having there fresh and very good water, with herbes of singular virtue. Not farre from hence we entred another Bay, where wee found people both men and women in their Canoas, naked, and ranging from one Island to another to seeke their meat, who entered traffique with us for such things as they had.

We returning hence Northward againe, found the 3 of October three Islands, in one of which was such plenty of birdes as is scant credible to report.

The 8 day of October we lost sight of one of our Consorts wherein M. Winter was, who as then we supposed was put by a storme into the streights againe, which at our returne home wee round to be true, and he not perished, as some of our company feared.

Thus being come into the height of The streights againe, we ran, supposing the coast of Chili to lie as the generall Maps have described it, namely Northwest, which we found to lie and trend to the Northeast and Eastwards, whereby it appeareth that this part of Chili hath not bene truly hitherto discovered, or at the least not truly reported for the space of 12 degrees at the least, being set downe either of purpose to deceive, or of ignorant conjecture.

We continuing our course, fell the 29 of November with an Island called la Mocha, were we cast anchor, and our Generall hoysing out our boate, went with ten of our company to shore, where wee found people, whom the cruell and extreme dealings of the Spaniards have forced for their owne safetie and libertie to flee from the maine, and to fortifie themselves in this Island. We being on land, the people came downe to us to the water side with shew of great courtesie, bringing to us potatoes, rootes, and two very fat sheepe, which our Generall received and gave them other things for them, and had promised to have water there: but the next day repayring againe to the shore, and sending two men aland with barrels to fill water, the people taking them for Spaniards (to whom they use to shew no favour if they take them) layde violent hands on them, and as we thinke, slew them.

Out Generall seeing this, stayed here no longer, but wayed anchor, and set sayle towards the coast of Chili, and drawing towards it, we mette neere to the shore an Indian in a Canoa, who thinking us to have bene Spaniards, came to us and tolde us, that at a place called S. Iago, there was a great Spanish ship laden from the kingdome of Peru: for which good newes our Generall gave him divers trifles, wherof he was glad, and went along with us and brought us to the place, which is called the port of Valparizo.

When we came thither, we found indeede the ship riding at anker, having in her eight Spaniards and three Negros, who thinking us to have bene Spaniards and their friends, welcomed us with a drumme, and made ready a Bottija of wine of Chili to drinke to us: but as soone as we were entred, one of our company called Thomas Moone began to lay about him, and stroke one of the Spanyards, and sayd unto him, Abaxo Perro, that is in English, Goe downe dogge. One of these Spaniards seeing persons of that quality in those seas, all to crossed, and blessed himselfe: but to be short, were stowed them under hatches all save one Spaniard, who suddenly and desperately leapt over boord into the sea, and swamme ashore to the towne of S. Iago, to give them warning of our arrivall.

They of the towne being not above 9 households, presently fled away and abandoned the towne. Our Generall manned his boate, and the Spanishships boate, and went to the Towne, and being come to it, we rifled it, and came to a small chappell which wee entred, and found therein a silver chalice, two cruets, and one altar-cloth, the spoyle whereof our Generall gave to M. Fletcher his minister.

We found also in this towne a warehouse stored with wine of Chili, and many boords of Cedar-wood, all which wine we brought away with us, and certaine of the boords to burne for fire-wood: and so being come aboord, wee departed the Haven, having first set all the Spaniards on land, saving one John Griego a Greeke borne whom our Generall carried with him for his Pilot to bring him into the haven of Lima.

When we were at sea, our Generall rifled the ship, and found in her good store of the wine of Chili, and 25000 pezoes of very pure and fine gold of Baldivia, amounting in value to 37000 ducats of Spanish money, and above. So going on our course, wee arrived next at a place called Coquimbo, where our Generall sent 14 of his men on land to fetch water: but they were espied by the Spaniards, who came with 300 horsemen and 200 footemen, and slewe one of our men with a piece, the rest came aboord in safetie, and the Spaniards departed: wee went on shore againe, and buried our man, and the Spaniards came downe againe with a flag of truce, but we set sayle and would not trust them.

From hence we went to a certaine port called Tarapaza, where being landed, we found by the sea side a Spaniard lying asleepe, who had lying by him 13 barres of silver, which weighed 4000 ducats Spanish; we tooke the silver, and left the man.

Not farre from hence going on land for fresh water, we met with a Spaniard and an Indian boy driving 8 llamas or sheepe of Peru which are as big as asses; every of which sheepe had on his backe 2 bags of leather, each bagge conteining 50 li weight of fine silver: so that bringing both the sheepe and their burthen to the ships, we found in all the bags 800 weight of silver.

Here hence we sailed to a place called Arica, and being entred the port, we found there three small barkes which we rifled, and found in one of them 57 wedges of silver, each of them weighing about 20 pound weight, and every of these wedges were of the fashion and bignesse of a brickbat. In all these 3 barkes we found not one person: for they mistrusting no strangers, were all gone aland to the Towne, which consisteth of about twentie houses, which we would have ransacked if our company had bene better and more in number. But our Generall contented with the spoyle of the ships, left the Towne and put off againe to sea and set sayle for Lima, and by the way met with a small barke, which he boorded, and found in her good store of linen cloth, whereof taking some quantitie, he let her goe.

To Lima we came the 13 day of February, and being entred the haven, we found there about twelve sayle of ships lying fast moored at an anker, having all their sayles carried on shore; for the masters and marchants were here most secure, having never bene assaulted by enemies, and at this time feared the approach of none such as we were. Our Generall rifled these ships, and found in one of them a chest full of royals of plate, and good store of silkes and linen cloth, and tooke the chest into his owne ship, and good store of the silkes and linen. In which ship hee had newes of another ship called the Cacafuego which was gone toward Paita, and that the same shippe was laden with treasure: whereupon we staied no longer here, but cutting all the cables of the shippes in the haven, we let them drive whither they would, either to sea or to the shore, and with all speede we followed the Cacafuego toward Paita, thinking there to have found her: but before we arrived there, she was gone from thence towards Panama, whom our Generall still pursued, and by the way met with a barke laden with ropes and tackle for ships, which hee boorded and searched, and found in her 80.li. weight of golde, and a crucifixe of gold with goodly great Emerauds set in it which he tooke, and some of the cordage also for his owne ship.

From hence we departed, still following the Cacafuego, and our Generall promised our company, that whosoever could first descrie her, should have his chaine of gold for his good newes. It fortuned that John Drake going up into the top, descried her about three of the clocke, and about six of the clocke we came to her and boorded her, and shotte at her three peeces of ordinance, and strake downe her Misen, and being entered, we found in her great riches, as jewels and precious stones, thirteene chests full of royals of plate, foure score pound weight of golde, and six and twentie tunne of silver. The place where we tooke this prize was called Cape de San Francisco, about 150 leagues from Panama.

The Pilots name of this Shipp was Francisco, and amongst other plate that our Generall found in this ship, he found two very faire guilt bowles of silver, which were the Pilots: to whom our Generall sayd: Senior Pilot, you have here two silver cups, but I must needes have one of them: which the Pilot because hee could not otherwise chuse, yielded unto, and gave the other to the steward of our Generalls ships.

When this Pilot departed from us, his boy sayde thus unto our Generall: Captaine, our ship shall be called no more the Cacafuego, but the Cacaplata, and your shippe shall be called the Cacafuego: which pretie speech of the Pilots boy ministred matter of laughter to us, both then and long after.

When our Generall had done what hee would with this Cacafuego hee cast her off, and we went on our course still towards the

West, and not long after met with a ship laden with linen and cloth and fine China-dishes of white earth, and great store of China-silks, of all which things we tooke as we listed.

The owner himselfe of this shipe was in her, who was a Spanish Gentleman, from whom our Generall tooke a Fawlcon of golde, with a great Emeraud in the breast thereof, and the Pilot of the ship he tooke also with him, and so cast the ship off.

This Pilot brought us to the haven of Guatulco, the towne where-of, as he told us, had but 17 Spaniards in it. As soone as we were entred this haven, we landed, and went presently to the towne, and to the Towne-house, where we found a Judge sitting in judgement, being associate with three other officers, upon three Negros that had conspired the burning of the Towne: both which Judges & prisoners we tooke and brought them a shipboard, and caused the chiefe Judge to write his letter to the Towne, to command all the Townesmen to avoid, that we might safely water there. Which being done, and they departed, we ransacked the Towne, and in one house we found a pot of the quantitie of a bushel, full of reals of plate, which we brought to our ship.

And here one Thomas Moone one of our company, tooke a Spanish Gentleman as hee was flying out of the towne, and searching him, he found a chaine of golde about him, and other jewels, which he tooke, and so let him goe.

At this place our Generall among other Spaniards, set ashore his Portugall Pilote, which hee tooke at the Islands of Cape Verde, out of a ship of S. Mary port of Portugall: and having set them ashore, we departed hence, and sailed to the island of Canno, where our Generall landed, and brought to shore his owne ship, and discharged her, mended, and graved her, and furnished our ship with water and wood sufficiently.

And while we were here, we espied a shippe, and set saile after her, and tooke her, and found in her two Pilots, and a Spanish Governour, going for the Islands of the Philippinas: we searched the shippe, and tooke some of her merchandizes, and so let her goe. Our Generall at this place and time, thinking himselfe both in respect of his private injuries received from the Spaniards, as also of their contempts and indignities offered to our countrey and Prince in generall, sufficiently satisfied, and revenged: and supposing that her Majestie at his returne would rest contented with this service, purposed to continue no longer upon the Spanish coasts, but began to consider and to consult of the best way for his Countrey.

He thought it not good to returne by the Streights, for two speciall causes: the one, lest the Spaniards should there waite, and attend for him in great number and strength, whose hands, hee being left but one ship, could not possibly escape. The other cause was the dangerous situation of the mouth of the streights in the South sea, where continuall stormes reigning and blustering, as he found by experience, besides the shoals and sands upon the coast, he thought it not a good course to adventure that way: he resolved therefore to avoyde these hazards, to goe forward to the Islandes of the Malucos, and therehence to saile the course of Portugals by the Cape of Buena Esperanza.

Upon this resolution, hee beganne to thinke of his best way to the Malucos, and finding himselfe where he now was becalmed, he saw that necessitie hee must be forced to take a Spanish course, namely to sayle somewhat Northerly to get a winde. We therefore set saile, and sayled 600 leagues at the least for a good winde, and thus much we sailed from the 16 of April, till the 3 of June.

The 5 day of June, being in 43 degrees towards the pole Arctike, we found the ayre so colde, that our men being grievously pinched with the same complained of the extremitie thereof, and the further we went, the more the colde increased upon us. Whereupon we thought it best for that time to seeke the land, and did so, finding it not mountainous, but low plaine land, till we came within 38 degrees towards the line. In which height it pleased God to send us into a faire and good Baye, with a good winde to enter the same.

In this Baye we anchored, and the people of the Countrey having their houses close by the waters side, shewed themselves unto us, and sent a present to our Generall.

When they came unto us, they greatly wondred at the things that we brought, but our Generall (according to his naturall and accustomed humanitie) courteously intreated them, and liberally bestowed on them necessary things to cover their nakednesse, whereupon they supposed us to be gods, and would not be perswaded to the contrary: the presents which they sent to our Generall were feathers, and calles of net-worke.

Their houses are digged round about with earth, and have from the uttermost brimmes of the circle, clifts of wood set upon them, joining close together at the toppe like a spire steeple, which by reason of that closenesse are very warme.

Their beds is the ground with rushes strowed on it, and lying about the house, have the fire in the midst. The men go naked, the women take bulrushes, and kembe them after the manner of hempe, and thereof make their loose garments, which being knit about their middles, hang down about their hippes, having also about their shoulders a skinne of Deere, with the haire upon it. These women are very obedient and serviceable to their husbands.

After they were departed from us, they came and visited us the second time and brought with them feathers and bags of Tabacco for presents: And when they came to the top of the hill (at the bottome whereof we had pitched our tents) they staied themselves: where one appointed for speaker wearied himselfe with making a long oration, which done, they left their bowes upon the hill, and came downe with their presents.

In the meane time the women remaining on the hill, tormented themselves lamentably, tearing their flesh from their cheeks, whereby we perceived that they were about a sacrifice. In the meane time our Generall with his company went to prayer, and to reading of the Scriptures, at which exercise they were attentive, & seemed greatly to be affected with it: but when they were come unto us, they restored againe unto us those things which before we bestowed upon them.

The newes of our being there being spread through the Countrey, the people that inhabited round about came downe, and amongst them the King himselfe, a man of a goodly stature, & comely personage, with many other tall and warlike men: before whose comming were sent two Ambassadors to our Generall, to signifie that their King was comming, in doing of which message, their speach was continued about halfe an houre. This ended, they by signes requested our Generall to send some thing by their hand to their King, as a token that his comming might be in peace: wherein our Generall having satisfied them, they returned with glad tidings to their King, who marched to us with a princely majestie, the people crying continually after their manner, and as they drew neere unto us, so did they strive to behave themselves in their action with comelinesse.

In the fore-front was a man of a goodly personage, who bare the scepter or mace before the King, whereupon hanged two crownes, a lesse and a bigger, with three chaines of a marveilous length: the crownes were made of knit worke wrought artificially with fethers of divers colours: the chaines were made of a bonie substance, and few be the persons among them that are admitted to weare them: and of that number also the persons are stinted, as some ten, some 12 &c.

Next unto him which bare the scepter, was the King himselfe, with his Guard about his person, clad with Conie skins, & other skins: after them followed the naked common sort of people, every one having his face painted, some with white, some with blacke, and other colours, & having in their hands one thing or another for a present, not so much as their children, but they also brought their presents.

In the meane time our Generall gathered his men together, and marched within his fenced place, making against their approching a very warre-like shew. They being trooped together in their order, and a generall salutation being made, there was presently a generall silence. Then he that bare the scepter before the King, being informed by another, whom they assigned to that office, with a manly and loftie voyce proclaymed that which the other spake to him in secrete, continuing halfe an houre: which ended, and a generall as it were given, the King with the whole number of men and women (the children excepted) came downe without any weapon, who descending to the foote of the hill, set themselves in order.

In comming towards our bulwarks and tents, the scepter bearer began a song, observing his measures in a daunce, and that with a stately countenance, whom the King with his Guarde, and every degree of persons following, did in like maner sing and daunce, saving onely the women, which daunced & kept silence. The Generall permitted them to enter within our bulwarke, where they continued their song and daunce a reasonable time. When they had satisfied themselves, they made signes to our Generall to sit downe, to whom the King, and divers others made several orations, or rather supplications, that hee would take their province and kingdome into his hand, and become their King, making signes that they would resigne unto him their right and little of the whole land, and become his subjects. In which, to perswade us the better, the King and the rest, with one consent, and with great reverence, joyfully singing a song, did set the crowne upon his head, inriched his necke with all their chaines, and offred unto him many other things, honouring him by the name of Hioh, adding thereunto as it seemed, a signe of triumph: which thing our Generall thought not meete to reject, because he knew not what honour and profit it might be to our Countrey. Wherefore in the name, and to the use of her Majestie he tooke the scepter, crowne, and dignitie of the said Countrey into his hands, wishing that the riches & treasure thereof might so conveniently be transported to the inriching of her kingdom at home, as it aboundeth in ye same.

The common sorte of people leaving the King and his Guarde with our Generall, scattered themselves together with their sacrifices among our people, taking a diligent views of every person: and such as pleased their fancie, (which were the yongest) they inclosing them about offred their sacrifices unto them with lamentable weepiang, scratching, and tearing the flesh from their faces with their nailes, whereof issued abundance of blood. But wee used signes to them of disliking this, and stayed their hands from force, and directed them upwards to the living God, whom onely they ought to worship. They shewed unto us their wounds, and craved helpe of them at our hands, whereupon we gave them lotions, plaisters, and oyntments, agreeing to the state of their griefes, beseeching God, to cure their diseases. Every third day they brought their sacrifices unto us, until they understood our meaning, that we had no pleasure in them: yet they could not be long absent from us, but dayly frequented our company to the houre of our departure, which departure seemed so grievous unto them, that their joy was turned into sorow. They intreated us, that being absent we would remember them, and by stealth provided a sacrifice, which we misliked.

Our necessarie businesse being ended, our Generall with his company travailed up into the Countrey to their villages, where we found herdes of Deere by 1000 in a company, being most large, and fat of body.

We found the whole Countrey to be a warren of a strange kinde of Connies, their bodies in bignesse as be the Barbary Connies, their heads as the heads of ours, the feete of a Want, and the taile of a Rat being of great length: under her chinne is on either side a bag, into the which she gathereth her meate, when she hath filled her bellie abroad. The people eate their bodies, and make great accompt of their skinnes, for their Kings coate was made of them.

Our Generall called this Countrey Nova Ablion, and that for two causes: the one in respect of the white bankes and cliffs, which lie towards the sea: and the other, because it might have some affinitie with our Countrey in name, which sometime was so called.

There is no part of earth heere to be taken up, wherein there is not some probable shew of gold or silver.

At our departure hence our Generall set up a monument of our being there, as also of her Majesties right and title to the same, namely a plate, nailed upon a faire great poste, where-upon was ingraven her Majesties name, the day and yeere of our arrival there, with the free giving up of the province and people into her Majesties hands, together with her highnesse picture and armes, in a peece of six pence of current English money under the plate, whereunder was also written the name of our Generall.

It seemeth that the Spaniards hitherto had never bene in this part of the Countrey, neither did ever discover the land by many degrees, to the Southwards of this place.

After we had set saile from hence, wee continued without sight of land till the 13 day of October following, which day in the morning wee fell with certaine Islands 8 degrees to the Northward of the line, from which Islands came a great number of Canoas, having in some of them 4 in some 6 and in some also 14 men, bringing with them cocos, and other fruites. Their Canoas were hollow within, and cut with great arte and cunning, being very smooth within and without, and bearing a glasse as if it were a horne daintily burnished, having a prowe, and a sterne of one sort, yielding inward circle-wise, being of a great height, and full of certaine white shels for a braverie, and on each side of them lie out two peeces of timber about a yard and a halfe long, more or lesse, according to the smalnesse, or bignesse of the boate.

This people have the nether part of their eares cut into a round circle, hanging downe very lowe upon their cheeks, whereon they hang things of a reasonable weight. The nailes of their hands are an ynche long, their teeth are as blacke as pitch, and they renew them often, by eating of an herbe with a kinde of powder, which they always carrie about them in a cane for the same purpose.

Leaving this Island the night after we fell with it, the 18 of October, we lighted upon divers others, some whereof made a great shew of Inhabitants.

We continued our course by the Islands of Tagulada, Zelon, and Zewarra, being friends to the Portugals, the first whereof hath growing in it great store of Cinnamom.

The 14 of November we fell with the Islands of Malcuo, which day at night (having directed our course to runne with Tydore) in coasting along the Island of Mutyr, belonging to the King of Ternate, his Deputie or Vice-king seeing us at sea, came with his Canoa to us without all feare, and came aboord, and after some conference with our Generall, willed him in any wise to rune in with Ternate, and not with Tydore, assuring him that the King would be glad of his comming, and would be ready to doe what he would require, for which purpose he himselfe would that night be with the King, and tell him the newes, with whom if he once dealt, hee should finde that as he was a King, so his word should stand: adding further, that if he went to Tydore before Ternate, the King would have nothing to doe with us, because hee held the Portugall as his enemie: whereupon our Generall resolved to runne with Ternate, where the next morning early we came to anchor, at which time our Generall sent a messenger to the King with a velvet cloke for a present, and token of his comming to be in peace, and that he required nothing but traffique and exchange of marchandize, whereof he had good store, in such things as he wanted.

In the meane time the Vice-king had bene with the King according to his promise, signifying unto him what good things he might receive from us by traffique: whereby the King was mooved with great liking towards us, and sent to our Generall with speciall message, that hee should have what things he needed, and would require with peace and friendship, and moreover that hee would yeeld himselfe, and the right of his Island to bee at the pleasure and commandement of so famous a Prince as we served. In token whereof he sent of our Generall a signet, and within short time after came in his owne person, with boates, and Canoas to our ship, to bring her into a better and safer roade then she was in at present.

In the meane time, our Generalls messenger beeing come to the Court, was met by certaine noble personages with great solemnitie, and brought to the King, at whose hands hee was most friendly and graciously intertained.

The King purposing to come to our ship, sent before 4 great and large Canoas, in every one whereof were certaine of his greatest states that were about him, attired in white lawne of cloth of Calicut, having over their heads from the one ende of the Canoa to the other, a covering of thinne perfumed mats, borne up with a frame made of reedes for the same use, under which every one did sit in his order according to his dignitie, to keepe him from the heate of the Sunne, divers of whom beeing of good age and gravitie, did make an ancient and fatherly shew. There were also divers yong and comely men attired in white, as were the others: the rest were souldiers, which stood in comely order round about on both sides, without whom sate the rowers in certaine galleries, which being three on a side all along the Canoas, did lie from the side thereof three or foure yardes, one being orderly builded lower then another, in every of which galleries were the number of 4 score rowers.

These Canoas were furnished with warlike munition, every man for the most part having his sword and target, with his dagger, beside other weapons, as launces, calivers, darts, bowes and arrowes: also every Canoa had a small cast base mounted at the least one full yarde upon a stocke set upright.

Thus comming neere our shippe, in order they rowed about us, one after another, and passing by, did their homage with great solemnitie, the great personages beginning with great gravitie and fatherly countenances, signifying that ye King had sent them to conduct our ship into a better roade.

Soone after the King himselfe repaired, accompanied with 6 grave and ancient persons, who did their obeisance with marveilous humilitie. The King was a man of tall stature, and seemed to be much delighted with the sound of our musicke, to whom as also to his nobilitie, our Generall gave presents, wherewith they were passing well contented.

At length the King craved leave of our Generall to depart, promising the next day to come aboord, and in the meane time to send us such victuals, as were necessarie for our provision: so that the same night we received of them meale, which they call Sagu, made of the tops of certaine trees, tasting in the mouth like sowre curds, but melteth like sugar, whereof they make certaine cakes, which may be kept the space of ten yeeres, and yet then good to be eaten. We had of them store of rice, hennes, unperfect and liquid sugar, sugar canes, and a fruite which they call Figo, with store of cloves.

The King having promised to come aboord, brake his promise, but sent his brother to make his excuse, and to intreate our Generall to come on shoare, offring himself pawne aboord for his safe returne. Whereunto our Generall consented not, upon mislike conceived of the breach of his promise, the whole company also utterly refusing it. But to satisfie him, our Generall sent certaine of his Gentlemen to the Court, to accompany the King's brother, reserving the Vice-king for their safe returne. They were received of another brother of the Kings, and other states, and were conducted with great honour to the Castle. The place that they were brought unto, was a large and faire house, where were at the least 1000 persons assembled.

The King being yet absent, there sate in their places 60 grave personages, all which were said to be of the Kings Counsel. There were besides 4 grave persons, apparelled all in red, downe to the ground, and attired on their heads like the Turkes, and these were said to be Romanes, and Ligiers there to keepe continual traffike with the people of Ternate. There were also 2 Turks Ligiers in this place, and one Italian. The King at last came in guarded with 12 launces covered over with a rich canopy, with embossed gold. Our men accompanied with one of their Captaines called Moro, rising to meete him, he graciously did welcome, and intertaine them. He was attired after the manner of the Countrey, but more sumptuously then the rest. From his waste downe to the ground, was all cloth of golde, and the same very rich: his legges were bare, but on his feete were a paire of Shooes, made of Cordovan skinne. In the attire of his head were finely wreathed hooped rings of gold, and about his necke he had a chaine of perfect golde, the linkes whereof were great, and one folde double. On his fingers hee had sixe very faire jewels, and sitting in his chaire of estate, at his right hand stood a page with a fanne in his hand, breathing and gathering the ayre to the King. The fanne was in length two foote, and in bredth one foote, set with 8 saphyres, richly imbrodered, and knit to a staffe 3 foote in length, by the which the Page did hold, and moove it. Our Gentlemen having delivered their message, and received order accordingly, were licensed to depart, being safely conducted backe againe by one of the Kings Counsell.

An Indonesian beach similar to ones that Drake and his crew probably saw on their travels.

This Island is the chiefest of all the Islands of Maluco, and the King hereof is King of 70 Islands besides. The King with his people are Moores in religion, observing certaine new Moones with fastings: during which fasts, they neither eat nor drinke in the day, but in the night.

After that our Gentlemen were returned, and that we had heere by the favour of the King received all necessary things that the place could yeeld us: our Generall considering the great distance, and how farre he was yet off from his Countrey, thought it not best here to linger the time any longer, but waying his anchors, set out of the Island, and sayled to a certaine little Island to the Southwards of Celebes, where we graved our ship, and continued there in that and other businesses 26 dayes. This Island is throughly growen with wood of a large and high growth, very straight and without boughes, save onely in the head or top, whose leaves are not much differing from our broome in England. Amongst these trees night by night, through the whole land, did shew themselves and infinite swarme of fiery wormes flying in the ayre, whose bodies beeing no bigger then our common English flies, make such a shew and light, as if every twigge or tree had bene a burning candle. In this place breedeth also wonderfull store of Bats, as bigge as large hennes: of Crayfishes also heere wanted no plentie, and they of exceeding bignesse, one whereof was sufficient for 4 hungry stomacks at a dinner, beeing also very good, and restoring meate, whereof we had experience: and they digge themselves holes in the earth like Conies.

When we had ended our businesse here, we waied, and set saile to runne for the Malucos: but having at that time a bad winde, and being amongst the Islands, with much difficultie we recovered to the Northward of the Island of Celebes, where by reason of contrary winds not able to continue our course to runne Westwards, we were inforced to alter the same to the Southward againe, finding that course also to be very hard and dangerous for us, by reason of infinite shoalds which lie off, and among the Islands: whereof we had too much triall to the hazard and danger of our shippe and lives. For of all other dayes upon the 9 of Januarie, in the yeere 1579 we ranne suddenly upon a rocke, where we stucke fast from 8 of the clocke at night, til 4 of the clocke in the afternoone the next day being indeede out of all hope to escape the danger: but our Generall as hee had alwayes hitherto shewed himselfe couragious, and of a good confidence in the mercie and protection of God: so now he continued in the same, and lest he should seeme to perish wilfully, both he, and we did our best indevour to save our selves, which it pleased God so to blesse, that in the ende we cleared our selves most happily of the danger.

We lighted our ship upon the rockes of 3 tunne of cloves, 8 peeces of ordinance, and certaine meale and beanes: and then the winde (as it were in a moment by the speciall grace of God) changing from the starreboord to the larboord of the ship, we hoised our sailes, and the happy gale drove our ship off the rocke into the sea againe, to the no litle comfort of all our hearts, for which we gave God such prayse and thanks, as so great a benefite required.

The 8 of Februarie following, we fell with the fruitfull Island of Barateve, having in the meane time suffered many dangers by windes and shoalds. The people of this Islands are comely in body and stature, and of a civil behaviour, just in dealing, and courteous to strangers, whereof we had the experience sundry wayes, they being most glad of our presence, and very ready to releeve our wants in those things which their Countrey did yeelde. The men goe naked, saving their heads and privities, every man having something or other hanging at their eares. Their women are covered from the middle downe to the foote, wearing a great number of bracelets upon their armes, for some had 8 upon each arme, being made some of bone, some of horne, and some of brasse, the lightest whereof by our estimation waied two ounces apeece.

With this people linnen-cloth is good marchandize, and of good request, whereof they make rols for their heads, and girdles to weare about them.

Their Island is both rich and fruitfull: rich in golde, silver, copper, and sulphur, wherein they seeme skilfull and expert, not onely to trie the same, but in working it also artificially into any forme and fashion that pleaseth them.

Their fruits be divers and plentifull, as nutmegs, ginger, long pepper, lemmons, cucumbers, cocos, figu, sagu, with divers other sorts: and among all the rest, wee had one fruite, in bignesse, forme, and huske, like a Bay berry, hard of substance, and pleasant of taste, which being sodden, becommeth soft, and is a most good and wholsome victuall, whereof we tooke reasonable store, as we did also of the other fruits and spices: so that to confesse a trueth, since the time that we first set out of our owne Countrey of England, we happened upon no place (Ternate onely excepted) wherein we found more comforts and better meanes of refreshing.

At our departure from Barateve, we set our course for Java major, where arriving, we found great courtesie, and honourable entertainment. This Island is governed by 5 Kings, whom they call Rajah: as Rajah Donaw, and Rajah Mang Bange, and Rajah Cabuccapollo, which live as having one spirite, and one minde.

Of these five we had foure a shipboord at once, and two or three often. They are wonderfully delighted in coloured clothes, as red and greene: their upper parts of their bodies are naked, save their heads, whereupon they weare a Turkish roll, as do the Maluccians: from the middle downward they weare a pintado silke, trailing upon the ground, in colour as they best like.

The Maluccians hate that their women should be seene of strangers: but these offer them of high courtesie, yea the Kings themselves.

The people are of goodly stature, and warlike, well provided of swords and targets, with daggers, all being of their owne worke, and most artificially done, both in tempering their mettall, as also in the forme, whereof we bought reasonable store.

They have an house in every village for their common assembly: every day they meete twise, men, women, and children, bringing with them such victuals as they thinke good, some fruites, some rice boiled, some hennes roasted, some sagu, having a table made 3 foote from the ground, whereon they set their meate, that every person sitting at the table may eate, one rejoycing in the company of another.

They boile their rice in an earthen pot, made in forme of a sugar loafe, being ful of holes, as our pots which we water our gardens withall, and it is open at the great ende, wherein they put their rice dried, without any moisture. In the meane time they have ready another great earthen pot, set fast in a fornace, boiling full of water, whereinto they put their pot with rice, by such measure, that they swelling become soft at the first, and by their swelling stopping the holes of the pot, admit no more water to enter, but the more they are boiled, the harder and more firme substance they become, so that in the end they are a firme & good bread, of the which with oyle, butter, sugar, and other spices, they make divers sorts of meates very pleasant of taste, and nourishing to nature.

The French pocks is here very common to all, and they helpe themselves, sitting naked from ten to two in the Sunne, whereby the venomous humour is drawen out. Not long before our departure, they tolde us, that not farre off there were such great Ships as ours, wishing us to beware: upon this our Captaine would stay no longer.

From Java Major we sailed for the cape of Good Hope, which was the first land we fell withall: neither did we touch with it, or any other land, untill we came to Sierra Leona, upon the coast of Guinea: notwithstanding we ranne hard aboord the Cape, finding the report of the Portugals to be most false, who affirme, that it is the most dangerous Cape of the world, never without intolerable stormes and present danger to traveilers, which come neere the same.

This Cape is a most stately thing, and the fairest Cape we saw in the whole circumference of the earth, and we passed by it the 18 of June.

From thence we continued our course to Sierra Leona, on the coast of Guinea, where we arrived the 22 of July, and found necessarie Provisions, great store of Elephants, Oisters upon trees of one kind, spawning and increasing infinitely, the Oister suffering no budde to grow. We departed thence the 24 day.

We arrived in England the third of November 1580 being the third yeere of our departure.

CHAPTER 2

THE SINGULAR FATE OF THE BRIG *POLLY*

by Ralph D. Paine

An aerial view of a New England harbor.

S team has not banished from the deep sea the ships that lift tall spires of canvas to win their way from port to port. The gleam of their topsails recalls the centuries in which men wrought with stubborn courage to fashion fabrics of wood and cordage that should survive the enmity of the implacable ocean and make the winds obedient. Their genius was unsung, their hard toil forgotten, but with each generation the sailing ship became nobler and more enduring, until it was a perfect thing. Its great days live in memory with a peculiar atmosphere of romance. Its humming shrouds were vibrant with the eternal call of the sea, and in a phantom fleet pass the towering East Indiaman, the hard-driven Atlantic packet, and the gracious clipper that fled before the Southern trades.

A hundred years ago every bay and inlet of the New England coast was building ships which fared bravely forth to the West Indies, to the roadsteads of Europe, to the mysterious havens of the Far East. They sailed in peril of pirate and privateer, and fought these rascals as sturdily as they battled with wicked weather. Coasts were unlighted, the seas uncharted, and navigation was mostly by guesswork, but these seamen were the flower of an American merchant marine whose deeds are heroic in the nation's story. Great hearts in little ships, they dared and suffered with simple, uncomplaining fortitude. Shipwreck was an incident, and to be adrift in lonely seas or cast upon a barbarous shore was sadly commonplace. They lived the stuff that made fiction after they were gone.

Your fancy may be able to picture the brig *Polly* as she steered down Boston harbor in December, 1811, bound to Santa Cruz with lumber and salted provisions for the slaves of the sugar plantations. She was only a hundred and thirty tons burden and perhaps eighty feet long. Rather clumsy to look at and roughly built was the *Polly* as compared with the larger ships that brought home the China tea and silks to the warehouses of Salem. Such a ship was a community venture. The blacksmith, the rigger, and the calker took their pay in shares, or 'pieces.' They became part owners, as did likewise the merchant who supplied stores and material; and when the brig was afloat, the master, the mate, and even the seamen were allowed cargo space for commodities that they might buy and sell to their own advantage. A voyage directly concerned a whole neighbourhood.

Every coastwise village had a row of keel-blocks sloping to the tide. In winter weather too rough for fishing, when the farms lay idle, the Yankee Jack-of-all-trades plied with his axe and adze to shape the timbers and peg together such a little vessel as the *Polly*, in which to trade to London or Cadiz or the Windward Islands. Hampered by an unfriendly climate, hard put to it to grow sufficient food, with land immensely difficult to clear, the New Englander was between the devil and the deep sea, and he, sagaciously, chose the latter. Elsewhere, in the early days, the forest was an enemy to be destroyed with great pains. The pioneers of Massachusetts, New Hampshire, and Maine regarded it with favour as the stuff with which to make stout ships and the straight masts they 'stepped' in them.

Nowadays, such a little craft as the *Polly* would be rigged as a schooner. The brig is obsolete, along with the quaint array of scows, ketches, pinks, brigantines and sloops which once filled the harbours and hove their hempen cables short to the clank of the windlass or capstan-pawl, while the brisk seamen sang a chanty to help the work along. The *Polly* had yards on both masts, and it was a bitter task to lay out in a gale of wind and reef the unwieldly single topsails. She would try for no record passages, but jogged sedately, and snugged down when the weather threatened.

On this tragic voyage she carried a small crew, Captain W. L. Cazneau, a mate, four sailors, and a cook who was a native Indian. No mention is to be found of any ill omens that forecasted disaster, such as a black cat, or a crossed-eyed Finn in the forecastle. Two passengers were on board, "Mr. J. S. Hunt and a negro girl nine years old." We know nothing whatever about Mr. Hunt, who may have been engaged in some trading 'adventure' of his own. Perhaps his kinsfolk had waved him a fare-ye-well from the pier-head when the *Polly* warped out of her berth.

The lone piccaninny is more intriguing. She appeals to the imagination and inspires conjecture. Was she a waif of the slave traffic whom some benevolent merchant of Boston was sending to Santa Cruz to find a home beneath kindlier skies? Had she been entrusted to the care of Mr. Hunt? She is unexplained, a pitiful atom visible for an instant on the tide of human destiny. She amused the sailors, no doubt, and that austere, copper-hued cook may have unbent to give her a doughnut when she grinned at the galley-door.

Four days out from Boston, on December 15, the *Polly* had cleared the perilous sands of Cape Cod and the hidden shoals of the Georges. Mariners were profoundly grateful when they had safely worked off shore in the wintertime and were past Cape Cod, which bore a very evil repute in those days of square-rigged vessels. Captain Cazneau could recall that sombre day of 1802 when three fine ships, the *Ulysses, Brutus,* and *Volusia,* sailing together from Salem for European ports, were wrecked next day on Cape Cod. The fate of those who were washed ashore alive was most melancholy. Several died of the cold, or were choked by the sand which covered them after they fell exhausted.

As in other regions where shipwrecks were common, some of the natives of Cape Cod regarded a ship on the beach as their rightful plunder. It was old Parson Lewis of Wellfleet, who, from his pulpit window, saw a vessel drive ashore on a stormy Sunday morning. "He closed his Bible, put on his outside garment, and descended from the pulpit, not explaining his intention until he was in the aisle, and then he cried out 'Start fair' and took to his legs. The congregation understood and chased pell-mell after him."

The brig *Polly* laid her course to the southward and sailed into the safer, milder waters of the Gulf Stream. The skipper's load of anxiety was lightened. He had not been sighted and molested by the British men-of-war that cruised off Boston and New York to hold up Yankee merchantmen and impress stout seamen. This grievance was to flame in a righteous war only a few months later. Many a voyage was ruined, and ships had to limp back to port short-handed, because their best men had been kidnapped to serve in British ships. It was an age when might was right on the sea.

The storm which overwhelmed the brig *Polly* came out of the south-east, when she was less than a week on the road to Santa Cruz. To be dismasted and waterlogged was no uncommon fate. It happens often nowadays, when little schooners creep along the coast, from Maine and Nova Scotia ports, and dare the winter blows to earn their bread. Men suffer in open boats, as has been the seafarer's hard lot for ages, and they drown with none to hear their cries, but they are seldom adrift more than a few days. The story of the *Polly* deserves to be rescued from oblivion because, so far as I am able to discover, it is unique in the spray-swept annals of maritime disaster.

Seamanship was helpless to ward off the attack of the storm that left the brig a sodden hulk. Courageously her crew shortened sail and made all secure when the sea and sky presaged a change of weather. There were no green hands, but men seasoned by the continual hazards of their calling. The wild gale smote them in the darkness of the night. They tried to heave to as the canvas was whirled away in fragments. The seams of the hull opened as she labored, and six feet of water flooded the hold. Leaking like a sieve, the *Polly* would never see port again.

Worse was to befall her. At midnight she was capsized, or thrown on her beam-ends, as the sailor's lingo has it. She lay on her side while the clamorous seas washed clean over her. The skipper, the mate, the four seamen, and the cook somehow clung to the rigging and grimly refused to be drowned. They were of the old breed, "every hair a rope-yarn and every finger a fish-hook." They even managed to find an axe and grope their way to the shrouds in the faint hope that the brig might right if the masts went overside. They hacked away, and came up to breathe now and then, until the foremast and mainmast fell with a crash, and the wreck rolled level. Then they slashed with their knives at the tangle of spars and ropes until they drifted clear. As the waves rush across a half-tide rock, so they broke over the shattered brig, but she no longer wallowed on her side.

At last the stormy daylight broke. The mariners had survived, and they looked to find their two passengers who had no other refuge than the cabin. Mr. Hunt was gone, blotted out with his affairs and his ambitions, whatever they were. The coloured child they had vainly tried to find in the night. When the sea boiled into the cabin and filled it, she had climbed to the skylight in the roof, and there she clung like a bat. They hauled her out through a splintered gap, and sought tenderly to shelter her in a corner of the streaming deck, but she lived no more than a few hours. It was better that this bit of human flotsam should flutter out in this way than to linger a little longer in this forlorn derelict of a ship. The *Polly* could not sink, but she drifted as a

mere bundle of boards with the ocean winds and currents, while seven men tenaciously fought off death and prayed for rescue.

The gale blew itself out, the sea rolled blue and gentle, and the wreck moved out into the Atlantic, having veered beyond the eastern edge of the Gulf Stream. There was raw salt pork and beef to eat, nothing else, barrels of which they fished out of the cargo. A keg of water which had been lashed to the quarter-deck was found to contain thirty gallons. This was all there was to drink, for the other water-casks had been smashed or carried away. The diet of meat pickled in brine aggravated the thirst of these castaways. For twelve days they chewed on this salty raw stuff, and then the Indian cook, Moho by name, actually succeeded in kindling a fire by rubbing two sticks together in some abstruse manner handed down by his ancestors. By splitting pine spars and a bit of oaken rail he was able to find in the heart of them wood which had not been dampened by the sea, and he sweated and grunted until the great deed was done. It was a trick which he was not at all sure of repeating unless the conditions were singularly favourable. Fortunately for the hapless crew of the *Polly*, their Puritan grandsires had failed in their amiable endeavour to extinguish the aborigine.

The tiny galley, or 'camboose' as they called it, was lashed to ring-bolts in the deck, and had not been washed into the sea when the brig was swept clean. So now they patched it up and got a blaze going in the brick oven. The meat could be boiled, and they ate it without stint, assuming that a hundred barrels of it remained in the hold. It had not been discovered that the stern-post of the vessel was staved in under water and all of the cargo excepting some of the lumber had floated out.

The cask of water was made to last eighteen days by serving out a quart a day to each man. Then an occasional rain-squall saved them for a little longer from perishing of thirst. At the end of the forty days they had come to the last morsel of salt meat. The *Polly* was following an aimless course to the eastward, drifting slowly under the influence of ocean winds and currents. These gave her also a southerly slant, so that she was caught by the vast movement of water which is known as the Gulf Stream Drift. It sets over toward the coast of Africa and sweeps into the Gulf of Guinea.

The derelict was moving away from the routes of trade to Europe into the almost trackless spaces beneath the tropic sun, where the sea glittered empty to the horizon. There was a remote chance that she might be described by a low-hulled slaver crowding for the West Indies under a might press of sail, with her human freightage jammed between decks to endure the unspeakable horrors of the Middle Passage. Although the oceans were populous with ships a hundred years ago, trade flowed on habitual routes. Moreover, a wreck might pass unseen two or three miles away. From the quarter-deck of a small sailing ship there was no such circle of vision as extends from the bridge of a steamer forty or sixty feet above the water, where the officers gaze through high-powered binoculars.

The crew of the *Polly* started at skies which yielded not the merciful gift of rain. They had strength to build them a sort of shelter of lumber, but whenever the weather was rough, they were drenched by the waves which played over the wreck. At the end of fifty days of this hardship and torment the seven were still alive, but then the mate, Mr. Paddock, languished and died. It surprised his companions, for, as the old record runs, he was a man of robust constitution who had spent his life in fishing on the Grand Banks, was accustomed to endure privations, and appeared the most capable of standing the shocks of misfortune of any of the crew. In the meridian of life, being about thirty-five years old, it was reasonable to suppose that, instead of the first, he would have been the last to fall a sacrifice to hunger and thirst and exposure, but Heaven ordered it otherwise.

Singularly enough, the next to go was a young seaman, spare and active, who was also a fisherman by trade. His name was Howe. He survived six days longer than the mate, and "likewise died delirious and in dreadful distress." Fleeting thunder-showers had come to save the others, and they had caught a large shark by means of a running bowline slipped over his tail while he nosed about the weedy hull. This they cut up and doled out for many days. It was certain, however, that unless they could obtain water to drink they would soon all be dead men on the *Polly*.

Captain Cazneau seems to have been a sailor of extraordinary resource and resolution. His was the unbreakable will to live and to endure which kept the vital spark flickering in his shipmates. Whenever there was strength enough among them, they groped in the water in the hold and cabin in the desperate hope of finding something to serve their needs. In this manner they salvaged an iron teakettle and one of the captain's flint-lock pistols. Instead of flinging them away, he sat down to cogitate, a gaunt, famished wraith of a man who had kept his wits and knew what to do with them.

At length he took an iron pot from the galley, turned the teakettle upside down on it, and found that the rims failed to fit together. Undismayed, the skipper whittled a wooden collar with a seaman's sheath-knife, and so joined the pot and the kettle. With strips of cloth and pitch scraped from the deck-beams, he was able to make a tight union where his round wooden frame set into the flaring rim of the pot. Then he knocked off the stock of the pistol and had the long barrel to use for a tube. This he rammed into the nozzle of the teakettle, and calked them as well as he could. The result was a crude apparatus for distilling seawater, when placed upon the bricked oven of the gallery.

Imagine those three surviving seamen and the stolid redskin of a cook watching the skipper while he methodically tinkered and puttered! It was absolutely the one and final chance of salvation. Their lips were black and cracked and swollen, their tongues lolled, and they could no more than wheeze when they tried to talk. There was now a less precarious way of making fire than by rubbing dry sticks together. This had failed them most of the time. The captain had saved the flint and steel from the stock of his pistol. There was tow or tarry oakum to be shredded fine and used for tinder. This smouldered and then burst into a tiny blaze when the sparks flew from the flint, and they knew that they would not lack the blessed boon of fire.

Together they lifted the precious contrivance of the pot and the kettle and tottered with it to the galley. There was an abundance of fuel from the lumber, which was hauled through a hatch and dried on deck. Soon the steam was gushing from the pistol-barrel, and they poured cool saltwater over the upturned spout of the teakettle to cause condensation. Fresh water trickled from the end of the pistol-barrel, and they caught it in a tin cup. It was scarcely more than a drop at a time, but they stoked the oven and lugged buckets of saltwater, watch and watch, by night and day. They roused in their sleep to go on with the task with the sort of dumb instinct. They were like wretched automatons.

So scanty was the allowance of water obtained that each man was limited to "four small wine glasses" a day, perhaps a pint. It was enough to permit them to live and suffer and hope. In the warm seas which now cradled the *Polly* the barnacles grew fast. The captain, the cook, and the three seamen scraped them off and for some time had no other food. They ate these shell-fish mostly raw, because cooking interfered with that tiny trickle of condensed water.

The faithful cook was the next of the five to succumb. He expired in March, after they had been three months adrift, and the manner of his death was quiet and dignified, as befitted one who might have been a painted warrior in an earlier day. The account says of him:

On the 15th of March, according to their computation, poor Moho gave up the ghost, evidently from want of water, though with much less distress than the others, and in full exercise of his reason. He very devoutly prayed and appeared perfectly resigned to the will of God who had so sorely afflicted him.

The story of the *Polly* is unstained by any horrid episode of cannibalism, which occurs now and then in the old chronicles of shipwrecks. In more than one seaport the people used to point at some weather-beaten mariner who was reputed to have eaten the flesh of a comrade. It made a marked man of him, he was shunned, and the unholy notoriety followed him to other ships and ports. The sailors of the *Polly* did cut off a leg of the poor, departed Moho, and used it as bait for sharks, and they actually caught a huge shark by so doing.

It was soon after this that they found the other pistol of the pair, and employed the barrel to increase the capacity of the still. By lengthening the tube attached to the spout of the teakettle, they gained more cooling surface for condensation, and the flow of fresh water now amounted to "eight junk bottles full" every twenty-four hours. Besides this, wooden gutters were hung at the eaves of the galley and of the rough shed in which they lived, and whenever rain fell, it ran into empty casks.

The crew was dwindling fast. In April, another seaman, Johnson by name, slipped his moorings and passed on to the haven of Fiddler's Green, where the souls of all dead mariners may sip their grog and spin their yarns and rest from the weariness of the sea. Three men were left aboard the *Polly*, the captain and two sailors.

The brig drifted into that fabled area of the Atlantic that is known as the Sargasso Sea, which extends between 16 degrees and 38 degrees North, between the Azores and the Antilles. Here the ocean currents are confused and seem to move in circles, with a great expanse of stagnant ocean, where the seaweed floats in tangled patches of red and brown and green. It was an old legend that ships once caught in the Sargasso Sea were unable to extricate themselves, and so rotted miserably and were never heard of again. Columbus knew better, for his caravels sailed through these broken carpets of weed, where the winds were so small and fitful that the Genoese sailors despaired of reaching anywhere. The myth persisted, and it was not dispelled until the age of steam. The doldrums of the Sargasso Sea were the dread of sailing ships.

The days and weeks of blazing calms in this strange wilderness of ocean mattered not to the blindly errant wreck of the *Polly*. She was a dead ship that had outwitted her destiny. She had no masts and sails to push her through these acres of leathery kelp and bright masses of weed which had drifted from the Gulf and the Caribbean to come to rest in this solitary, watery waste. And yet to the captain and his two seamen this dreaded Sargasso Sea was beneficent. The stagnant weed swarmed with fish and gaudy crabs and molluscs. Here was food to be had for the mere harvesting of it. They hauled masses of weed over the broken bulwarks and picked off the crabs by hundreds. Fishing gear was an easy problem for these handy sailor-men. They had found nails enough; hand-forged and malleable. In the galley they heated and hammered them to make fish-hooks, and the lines were of small stuff 'unrove' from a length of halyard. And so they caught fish, and cooked them while the oven could be spared. Otherwise they ate them raw, which was not distasteful after they had become accustomed to it. The natives of the Hawaiian Islands prefer their fish that way. Besides this, they split a large number of small fish and dried them in the hot sun upon the roof of the shelter. The sea-salt which collected in the bottom of the still was rubbed into the fish. It was a bitter condiment, but it helped to preserve them against spoiling.

The season of spring advanced until the derelict *Polly* had been four months afloat and wandering, and the end of the voyage was a long way off. The minds and bodies of the castaways had adjusted themselves to the intolerable situation. The most amazing aspect of the experience is that these men remained sane. They must have maintained a certain order and routine of distilling water, of catching fish, of keeping track of the indistinguishable procession of the days and weeks. Captain Cazneau's recollection was quite clear when he came to write down his account of what had happened. The one notable omission is the death of another sailor, name unknown, which must have occurred after April. The only seaman who survived to keep the skipper company was Samuel Badger.

By way of making the best of it, these two indomitable seafarers continued to work on their rough deckhouse, "which by constant improvement had become much more commodious." A few bundles of hewn shingles were discovered in the hold, and a keg of nails was found lodged in corner of the forecastle. The shelter was finally made tight and waterproof, but, alas! there was no need of having it "more commodious." It is obvious, also, that "when reduced to two only, they had a better supply of water." How long they remained in the Sargasso Sea it is impossible to ascertain. Late in April it is recounted that "no friendly breeze wafted to their side the sea-weed from which they could obtain crabs or insects."

The mysterious impulse of the currents plucked at the keel of the *Polly* and drew her clear of this region of calms and of ancient, fantastic sea-tales. She moved in the open Atlantic again, without guidance or destination, and yet she seemed inexplicably to be following an appointed course, as though fate decreed that she should find rescue waiting somewhere beyond the horizon.

The brig was drifting toward an ocean more frequented, where the Yankee ships bound out to the River Plate sailed in a long slant far over to the African coast to take advantage of the booming trade-winds. She was also wallowing in the direction of the route of the East India-men, which departed from English ports to make the far-distant voyage around the Cape of Good Hope. None of them sighted the speck of a derelict, which floated almost level with the sea and had no spars to make her visible. Captain Cazneau and his companion saw sails glimmer against the sky-line during the last thousand miles of drift, but they vanished like bits of cloud, and none passed near enough to bring salvation.

June found the *Polly* approaching the Canary Islands. The distance of her journey had been about two thousand miles, which would make the average rate of drift something more than three hundred miles a month, or ten miles per day. The season of spring and its apple blossoms had come and gone in New England, and the brig had long since been mourned as missing with all hands. It was on the 20th of June that skipper and his companion—two hairy, ragged apparitions—saw three ships which appeared to be heading in their direction. This was in latitude 28 degrees North and longitude 13 degrees West, and if you will look at a chart you will note that the wreck would soon have stranded on the coast of Africa. The three ships, in company, bore straight down at the pitiful little brig, which trailed fathoms of sea-growth along her hull. She must have seemed uncanny to those who beheld her and wondered at the living figures that moved upon the weather-scarred deck. She might have inspired "The Ancient Mariner."

Not one ship, but three, came bowling down to hail the derelict. They manned the braces and swung the main-yards aback, beautiful, tall ships and smartly handled, and presently they lay hove to. The captain of the nearest one shouted a hail through his brass trumpet, but the skipper of the *Polly* had no voice to answer back. He sat weeping upon the coaming of a hatch. Although not given to emotion, he would have told you that it had been a hard voyage. A boat was dropped from the davits of this nearest ship, which flew the red ensign from her spanker-gaff. A few minutes later Captain Casneau and Samuel Badger, able seaman, were alongside the good ship *Fame* of Hull, Captain Featherstone, and lusty arms pulled them up the ladder. It was six months to a day since the *Polly* had been thrown on her beam-ends and dismasted.

The three ships had been near together in light winds for several days, it seemed, and it occurred to their captains to dine together on board of the *Fame*. And so the three skippers were there to give the survivors of the *Polly* a welcome and to marvel at the yarn they spun. The *Fame* was homeward bound from Rio Janeiro. It is pleasant to learn that Captain Cazneau and Samuel Badger "were received by these humane Englishmen with expressions of the most exalted sensibility." The musty old narrative concludes:

Thus was ended the most shocking catastrophe which our seafaring history has recorded for many years, after a series of distresses from December 20 to the 20th of June, a period of one hundred and ninety-two days. Every attention was paid to the sufferers that generosity warmed with pity and fellow-feeling could dictate, on board the *Fame*. They were transferred from this ship to the brig *Dromio* and arrived in the United States in safety.

Here the curtain falls. I for one should like to hear more incidents of this astonishing cruise of the derelict *Polly* and also to know what happened to Captain Cazneau and Samuel Badger after they reached the port of Boston. Probably they went to sea again, and more than likely in a privateer to harry British merchantmen, for the recruiting officer was beating them up to the rendezvous with fife and drum, and in August of 1812 the frigate *Constitution*, with ruddy Captain Isaac Hull walking the poop in a gold-laced coat, was pounding the *Guerrière* to pieces in thirty minutes, with broadsides whose thunder echoed round the world.

"Ships are all right. It is the men in them," said one of Joseph Conrad's wise old mariners. This was supremely true of the little brig that endured and suffered so much, and among the humble heroes of blue water by no means the least worthy to be remembered are Captain Cazneau and Samuel Badger, able seaman, and Moho, the Indian cook.

CHAPTER 3

SHIPWRECK

by Captain James Riley

"The Storm on the Sea of Galilee" by Rembrandt Harmenszoon van Rijn, 1633.

We set sail from the bay of Gibraltar on the 23rd of August, 1815, intending to go by way of the Cape de Verde Islands, to complete the landing of the vessel with salt. We passed Capt Spartel on the morning of the 24th, giving it a berth of from ten to twelve leagues, and steered off to the W. S. W. I intended to make the Canary Islands, and pass between Teneriffe and Palma, having a fair wind; but it being very thick and foggy weather, though we got two observations at noon, neither could be much depended upon. On account of the fog, we saw no land, and found, by good meridian altitudes on the twenty-eighth, that we were in the latitude of N 27.30 having differed our latitude by the force of current, one hundred and twenty miles; thus passing the Canaries without seeing any of them. I concluded we must have passed through the intended passage without discovering the land on either side, particularly as it was in the night, which was very dark, and black as pitch; nor could I believe otherwise from having had a fair wind all the way, and having steered one course ever since we took our departure from Cape Spartel. Soon after we got an observation on the 28th, it became as thick as ever, and the darkness seemed (if possible) to increase. Towards evening I got up my reckoning, and examined it all over, to be sure that I had committed no error, and caused the mates to do the same with theirs. Having thus ascertained that I was correct in calculation, I altered our course to S.W. which ought to have carried us nearly on the course I wished to steer, that is, for the easternmost of the Cape de Verds; but finding the weather becoming more foggy towards night, it being so thick that we could scarcely see the end of the jib-boom, I rounded the vessel to, and sounded with one hundred and twenty fathoms of line, but found no bottom, and continued on our course, still reflecting on what should be the cause of our not seeing land. As I never had passed near the Canaries before without seeing them, even in thick weather or in the night, I came to a determination to haul off to the N.W. by the wind at 10 P.M., as I should then be by the long only thirty miles north of Cape Bajador. I concluded on this at nine, and thought my fears had never before so much prevailed over my judgment and my reckoning. I ordered the light sails to be handed, and the steering sail booms to be rigged in snug, which was done as fast as it could be by one watch, under the immediate direction of Mr. Savage.

We had just got the men stationed at the braces for hauling off, as the man at helm cried "ten o'clock." Out try-sail boom was on the starboard side, but ready for jibing; the helm was put to port, dreaming of no danger near. I had been on deck all the evening myself; the vessel was running at the rate of nine or ten knots, with a very strong breeze, and high sea, when the main boom was jibed over, and I at that instant heard a roaring; the yards were braced up—all hands were called. I imagined at first it was a squall, and was near ordering the sails to be lowered down; but I then discovered breakers foaming at a most dreadful rate under our lee. Hope for a moment flattered me that we could fetch off still, as there were no breakers in view ahead: the anchors were made ready; but these hopes vanished in an instant, as the vessel was carried by a current and a sea directly towards the breakers, and she struck! We let go the best bower anchor; all sails were taken in as fast as possible: surge after surge came thundering on, and drove her in spite of anchors, partly with her head on shore. She struck with such violence as to start every man from the deck. Knowing there was no possibility of saving her, and that she must very soon bilge and fill with water, I ordered all the provisions we could get at to be brought on deck, in hopes of saving some, and as much water to be drawn from the large casks as possible. We started several quarter casks of wine, and filled them with water. Every man worked as if his life depended upon his present exertions; all were obedient to every order I gave, and seemed perfectly calm. The vessel was stout and high, as she was only in ballast trim; the sea combed over her stern and swept her decks; but we managed to get the small boat in on deck, to sling her and keep her from staving. We cut away the bulwark on the larboard side so as to prevent the boats from staving when we should get them out; cleared away the long boat and hung her in tackles, the vessel continuing to strike very heavy, and filling fast. We, however, had secured five or six barrels of water, and as many of wine, three barrels of bread, and three or four salted provisions. I had as yet been so busily employed, that no pains had been taken to ascertain what distance we were from the land, nor had any of us yet seen it; and in the meantime all the clothing, chests, trunks, etc. were got up, and the books, charts, and sea instruments, were stowed in them, in the hope of their being useful to us in future.

The vessel being now nearly full of water, the surf making a fair breach over her, and fearing she would go to pieces, I prepared a rope, and put it in the small boat, having got a glimpse of the short, at no great distance, and taking Porter with me, we were lowered down on the larboard or lee side of the vessel, where she broke the violence of the sea, and made it comparatively smooth; we shoved off, but on clearing away from the bow of the vessel, the boat was overwhelmed with a surf, and we were plunged into the foaming surges: we were driven along by the current, aided by what seamen call the undertow, (or recoil of the sea) to the distance of three hundred yards to the westward, covered nearly all the time by the billows, which, following each other in quick succession, scarcely gave us time to catch a breath before we were again literally swallowed by them, till at length we were thrown, together with our boat, upon a sandy beach. After taking breath a little, and ridding our stomachs of the saltwater that had forced its way into them, my first care was to turn the water out of the boat, and haul her up out of the reach of the surf. We found the rope that was made fast to her still remaining; this we carried up along the beach, directly to leeward of the wreck, where we fastened it to sticks about the thickness of handspikes, that had drifted on the shoe from the vessel, and which we drove into the sand by the help of other pieces of wood. Before leaving the vessel, I had directed that all the chests, trunks, and everything that would float, should be hove overboard: this all hands were busied in doing. The vessel lay about one hundred fathoms from the beach, at high tide. In order to save the crew, a hawser was made fast to the rope we had on shore, one end of which we hauled to us, and made it fast to a number of sticks we had driven into the sand for the purpose. It was then tautened on board the wreck, and made fast. This being done, the long-boat (in order to save the provisions already in her) was lowered down, and two hands steadied her by ropes fastened to the rings in her stem and stern posts over the hawser, so as to slide, keeping her bow to the surf. In this manner they reached the beach, carried on the top of a heavy wave. The boat was stove by the violence of the shock against the beach; but by great exertions we saved the three barrels of bread in her before they were much damaged; and two barrels of salted provisions were also saved. We were now, four of us, on shore, and busied in picking up the clothing and other things which drifted from the vessel, and carrying them up out of the surf. It was by this time daylight, and high water; the vessel careened deep off shore, and I made signs to have the masts cut away, in the hope of easing her, that she might not go to pieces. They were accordingly cut away, and fell on her starboard side, making a better lee for a boat alongside the wreck, as they projected considerably beyond her bows. The masts and rigging being gone, the sea breaking very high over the wreck, and nothing left to hold on by, the mates and six men still on board, though secured, as well as they could be, on the bowsprit and in the larboard fore-channels, were yet in imminent danger of being washed off by every surge. The long-boat was stove, and it being impossible for the small one to live, my great object was now to save the lives of the crew by means of the hawser. I therefore made signs to them to come, one by one, on the hawser, which had been stretched taut for that purpose. John Hogan ventured first, and having pulled off his jacket, took to the hawser, and made for the shore. When he had got clear of the immediate lee of the wreck, every surf buried him, combing many feet above his head; but he still held fast to the rope with a death-like grasp, and as soon as the surf was passed, proceeded on towards the shore, until another surf, more powerful than the former, unclenched his hands, and threw him within our reach; when we laid hold of him and dragged him to the beach; we then rolled him on the sand, until he discharged the saltwater from his stomach, and revived. I kept in the water up to my chin, steadying myself by the hawser, while the surf passed over me, to catch the others as they approached, and thus, with the assistance of those already on shore, was enabled to save all the rest from a watery grave.

CHAPTER 4

THE FAMOUS PERRY EXPEDITION

by John S. Sewall

Commodore Matthew C. Perry, USN. Credit: Matthew Brady

A t last, on the second of July 1853, four of the fleet got underway for Japan. The Saratoga took her place in tow of the Susquehanna as before, and the Plymouth in tow of the Mississippi. The Supply storeship was left for the time at anchor in Napa harbor, and the Caprice, under command of Lieutenant William L.

Maury, was sent to Shanghai. Our course followed the chain of island groups that extend to the northward and eastward from Lew Chew over to Nippon— some of the time in sight of them. One of the last we passed was Ohosima, a well-bred volcano that was enjoying a nice quiet smoke all by itself. It may wake up someday and start its furnaces, as its fiery neighbor Torisima has been doing while these pages have been in process of incubation. There is plenty of time; geology will furnish all it wants. And it may yet make its record in history and hold up its head with Vesuvius and Krakatoa and Mont Pelée. They are uncertain characters, these volcanoes; you can never be quite sure when any given island is preparing to burn out its chimney. The safest plan is to follow Confucius's advice about the gods—"Respect them, and keep out of their way."

We made moderate speed and reached Japan on the eighth of July. It was Friday, a memorable day in our calendar. That morning the lookouts at the masthead echoed through the fleet the rousing call, "Land ho!" We rushed on deck. There it was, at last. There it was, a dark silent cloud on the northern horizon, a terra incognita still shrouded in mystery, still inspiring the imagination with an indefinable awe, just as it had years ago in the studies of our childhood at school. We came up with it rapidly. But the rugged headlands and capes still veiled themselves in mist, as if resolved upon secrecy to the last. About noon the fog melted away, and there lay spread before us the Empire of the Rising Sun, a living picture of hills and valleys, of fields and hedges, groves, orchards, and forests that tufted the lawns and mantled the heights, villages with streets just a trifle wider, and houses a little less densely packed than those in China, defended by forts mounted with howitzers and "quakers," and fenced with long strips of black and white cotton, which signified that the fortifications were garrisoned and ready for business. On the waters were strange boats skimming about, impelled by strange boatmen, uncouth junks wafted slowly along by the breeze, vanishing behind the promontories and reappearing in the distance, or lowering their sails and dropping their four-fluked anchors in the harbours near us. And towering above all, forty miles inland, like a giant man-at-arms standing sentry over the scene, rose the snowy peak of Fusiyama, an extinct volcano fourteen thousand feet high, one of the most shapely cones in the world and well named "the matchless mountain."

Our squadron comprised, as already noted, two steam frigates and two sloops of war. For equipment we mustered sixty-one guns and 977 officers and men—a respectable force for the times, but soon eclipsed and forgotten in the vaster armaments of the Civil War and of our late scrimmage with Spain. Such a warlike apparition in the bay, small as it was, created a powerful sensation. A Japanese writer informs us that "the popular commotion in Yedo . . . was beyond description. The whole city was in an uproar. In all directions were seen mothers flying with children in their arms and men with mothers on their backs. Rumors of an immediate action, exaggerated each time they were communicated from mouth to mouth, added horror to the horror-stricken. The tramp of war-horses, the clatter of armed warriors, the noise of carts, the parade of firemen, the incessant tolling of bells, the shrieks of women, the cries of children, dinning all the streets of a city of more than a million souls, made confusion worse confounded.

Of all this we were quite unconscious. We had no idea that we had frightened the empire so badly, the capital being some forty or fifty miles away from our anchorage. But that the town near us was thrown into convulsions by the big "black fireships of the barbarians," as the Japanese called us, was sufficiently evident. Before our anchors were fairly down, a battery on Cape Kamisaki sent a trio of bomshells to inquire after our health, or perhaps to consign us to perdition. But they exploded harmlessly astern, and we sent no bombshells back to explain how we were, or whether we intended going in the direction indicated. Our friends on shore knew something of guns and gunnery—that was plain. How much, we could not tell. But our glasses showed us that not all the black logs frowning at us from their portholes were genuine. Some at least were "quakers," that could not be fired except in a general conflagration; like the battery of a native guard boat in the harbour of Nagasaki that once upon a time capsized in a squall; various things went to the bottom, but most of her guns floated!

By the time we were well anchored and sails furled and men piped down, swarms of picturesque Mandarins came off to challenge the strange arrival and to draw around the fleet the customary cordon of guard boats. This looked like being in custody. The American ambassador had not come to Japan to be put under sentries. He notified the Mandarins that his vessels were not pirates and need not be watched. They pleaded Japanese law. He replied with American law. They still insisted. Whereupon he clinched the American side of the argument with the notice that if the boats were not off in fifteen minutes, he should be obliged to open his batteries and sink them. That was entirely convincing, and the guard boats stood not on the order of their going but betook themselves to the shelter of the stone.

I well remember that still starlit night that closed our first day in Yedo Bay. Nothing disturbed its peaceful beauty. The towering ships slept motionless on the water, and the twinkling lights of the towns along the shore went out one by one. A few beacon fires lighted upon the hilltops, the rattling cordage of an occasional passing junk, the musical tones of a distant temple bell that came rippling over the bay at intervals throughout the long night— these were to us the only tokens of life in the sleeping empire.

A sleeping empire truly; aloof from the world, shut in within itself and utterly severed from the general world-consciousness, not awake to the opportunities and privileges it was later so suddenly and so brilliantly to achieve as one of the world's powers, not even conscious that there was any such high position to be attained. While the expedition is resting its anchors, and the empire around is asleep, let us take the chance to paint in a bit of the background. An historical reminiscence or two will enable us more fully to appreciate the aim and the ultimate success of the enterprise.

The Sunrise Kingdom, like the telescope, was discovered by accident. In 1542, when Henry VIII of England, Charles V of Germany, Knox, Calvin, and Luther were the chief characters on the European stage, a Portuguese vessel bound to Macao in China was driven by

United States East India Squadron in Tokyo Bay by Osaki, 1870s.

storms into Bungo, a port of Kiusiu. It was the first meeting of Japanese and Europeans. It seems to have been mutually agreeable. The accidental visitors were dazzled with the riches of the oriental paradise they had found, and the natives were pleased and entertained with their outlandish guests. When the news reached Europe, it started a crusade of adventurers to the eastern seas. There was gold fever; all the commercial nations of the West had even caught it. The flags of Portugall, England, Holland, France, and Spain soon waved in succession over the waters of the newly discovered empire. The Japanese were amiable, and a busy barter was maintained for some scores of years.

Traders and speculators were not the only visitors in that distant mart. Some ten years later the Jesuits resolved to signalize the beginnings of their new order by converting those rich and dissolute Gentiles. Their crusade, like many others, was successful. It is related that when the first missioners, as they were called, reached the field of their operations, some of the countries desired an edict against the propagation of the new faith. "How many religions have we now?" asked the emperor. "Thirty-five," was the answer. "Very well," said the tolerant monarch, "One more will hurt nobody—let them preach." And they did preach. And Xavier, the renowned Jesuit apostle and saint, though within the year he returned to China to die, lived long enough to baptize multitudes of the penitent pagans, grandees as well as commoners and peasantry. Other missioners flocked to the harvest. The Jesuits were then followed by Dominicans and Franciscans. The splendid robes and ritual of the church proved attractive and large numbers of the people were gathered into the Roman fold. Shrines were deserted and priests found their customs wasting away.

This was a result not entirely palatable to either the priesthood or the court. Several of the emperors recalled their apostate subjects to the mourning gods. Persecutions began. The foreign monks and friars were accused of political intrigue. The story is a bloody one and covers a whole generation of tragedy and horror. Let us turn the page and simply record the fact that Christianity was expunged from Japan.

The final catastrophe occurred in 1637 at the fall of Simabara and the massacre of some forty thousand Christians. The histories tell us that the bodies of the martyrs were tumbled together into one vast pit and over it was raised this defiant inscription: "So long as the sun shall warm the earth let no Christian be so bold as to come to Japan; and let all know that the King of Spain himself, or the Christian's God, or the most great God of all, if he violate this command, shall pay for it with his head." Then the murderous empire wiped its sword, shut its gates, and barred itself in against all the world. One of the precautions by which it protected itself against Christianity and the civilization of the West was the famous ceremony of trampling on the cross; the astute pagans rightly divining that no foreigner would consent to such a sacrilege who had enough of the Christian religion about him to disturb the empire.

The ceremony was performed every year, as methodically as taking the census or collecting the taxes, and was only abolished as late as 1853, after our first visit to Japan. Once a year, officers went to every house with boxes containing the crucifix and images of the Virgin. These were laid on the floor, and all the household from octogenarians to infants in arms were required to tread upon them as a proof that they were not Christians. This law was enforced among the Dutch, the only western nation that maintained its foothold in the hermit land during all those darkened centuries. It is said that the cross was carved into the stone thresholds of their warehouses so that they could neither go nor come without trampling upon it. The placid Hollanders do not seem to have been much distressed by the requirement; their convenient religion was easily detached and left in Europe. One of them, we are told, one day wandered away from the warehouses on the island of Dezima across the bridge into the streets of Nagasaki and was suddenly halted by a Japanese patrol. "Are you a Christian?" was the challenge. "No, I am a Dutchman!" He was allowed to pass.

It is time to return to the ships. We left them sound asleep at anchor off Uraga the night of our arrival in Yedo Bay. Yet not all sound asleep, for a more vigilant watch has rarely been kept than was kept that night on board that fleet. Nothing happened however—except a brilliant display of meteoric light in the sky during the midwatch, an omen that terribly alarmed our friends on shore as portending that the very heavens themselves were enlisted on the side of these foreign barbarians. The commodore alludes to the phenomenon in his narrative and adds the devout wish, "The ancients would have construed this remarkable appearance of the heavens as a favourable omen for any enterprise they had undertaken; it may be so construed by us, as we pray God that our present attempt to bring a singular and isolated people into the family of civilized nations may succeed without resort to bloodshed. In spite of the menacing sky we all survived, Yankees and natives, and in the morning were all alive and ready for business.

During the day, our new friends came off to visit the ships and some were admitted on board. These first interviews were a constant surprise to us; we found them so well-informed. They questioned us about the Mexican War, then recent; about General Taylor and General Santa Anna. On board the Susquehanna one day, a Japanese gentleman asked the officer of the deck, "Where did you come from?" "From America," the officer replied. "Yes, I know," he said, "Your whole fleet came from the United States. But this ship—did she come from New York? or Philadelphia? or Washington?" He knew enough of our geography not to locate our seaports on our western prairies or up among the Rockies—a pitch of intelligence not yet too common among even our European friends. One of them asked if the monster gun on the quarter-deck was a "paixhan" gun? Yes, it was, but where and how could he ever have heard the name? When two or three midshipmen were taking the sun at noon, one of them laid his sextant down and a Japanese taking it up remarked that such instruments came from London and Paris and the best were made in London. How could a Japanese know that?

Our colloquies were carried on in Dutch through our Dutch interpreter, Mr. Portman, the educated Japanese being then accustomed to the use of that language somewhat as we use French. We naturally supposed, therefore, that all their information had come through the Dutch, the only nation beside the neighboring Chinese and Koreans that had for the last centuries kept its hold upon the good graces and the commerce of Japan. But we afterwards found that the Japanese printers were in the habit of republishing the textbooks prepared by our missionaries in China for use in their schools. The knowledge of America which we found thus diffused in Japan had come straight from Dr. Bridgeman's History of the United States, a manual written and published in China, which also had, what the good doctor never dreamed of, a wide circulation in the realm of the Mikado. That book had already prepossessed its readers in our favour. The following winter it was my privilege to make the acquaintance of the author at his home in Shanghai and to sit often at his genial board. It has been one of the regrets of my life that I could not tell him and his accomplished wife that his little textbook was speeding its way within the Empire of the Rising Sun. But at that time, alas, none of us knew it. They have both long since gone home to the heaven they loved and probably never learned in this world of the good they had thus unconsciously done.

The next day was Sunday. According to custom, divine service was held on board the flagship. The capstan on the quarterdeck was draped with the flag and the Bible was laid open upon it. Chaplain Jones took his station beside it. I do not know that any record was made of the service; presumably the chaplain followed the usual liturgical form and preached a brief sermon. But the hymn sung on the occasion has become historic; it was Watts' solemn lyric:

> Before Jehovah's awful throne,
> Ye nations, bow with sacred joy.

It was sung to the tune of "Old Hundred" and was led by the full band. The familiar strains poured in mighty chorus from two hundred or three hundred lusty throats with a peal that echoed through the fleet and wafted the gracious message to the distant shore. The Japanese listened with wonder; and their wonder deepened into amazement when they found that the whole day was to be observed as a day of rest and none of them could be admitted on board.

On Monday the secular tide was turned on again and diplomatic overtures began in good earnest. In their official dealings with us it was interesting to see how the authorities clung to their time-honored policy of exclusion. It was a curious contest of steady nerve on one side, met by the most nimble parrying on the other. First they directed the commodore to go home; they wanted no letters from American presidents, nor any treaty. But the commodore would not go home. Then they ordered him to Nagasaki, where foreign business could be properly transacted through the Dutch. But the commodore declined to go to Nagasaki. If then this preposterous barbarian would not budge, and his letter must be received, they would receive it without ceremony on board ship. But his Western mightiness would not deliver it on board ship. Then they asked for time to consult the court at Yedo, and the commodore gave them three days—days big with fate; but exactly what happened at court we may never know. This much is certain, that our reluctant friends yielded at last; that pestilent letter would be received, and commissioners of suitable rank would come from court for the purpose. Even after all preliminaries had been settled, they begged to receive the letter on board ship, not on shore. But the Rubicon had been crossed.

Some three miles below our anchorage a little semicircular harbor makes in on the western side of the bay, and at the head of it stands the village or hamlet of Kurihama. That was the spot selected for the meeting of the Western envoy and the imperial commissioners, and there the Japanese erected a temporary hall of audience. It was a memorable scene. The two frigates steamed slowly down and anchored off the harbor. How big, black, and sullen they looked, masterful, accustomed to having their own way, full of pent-up force. Our little flotilla of fifteen boats landed under cover of their guns. We were not quite three hundred all told, but well befeathered in full uniform and armed to the teeth; a somewhat impressive lot, and yet of rather scant dimensions to confront five thousand native troops drawn up on the beach to receive us, with crowds of curious spectators lining the housetops and grouped on the hills in the rear. However, we were ready for anything and had no fear of treachery. The emblazonry of those Japanese regiments surpasses any powers of description that have been vouchsafed to the present deponent. Their radiant uniforms and trappings and ensigns must have been cut out of rainbows and sunsets; and the scores of boats fringing the shore heightened the effect with their fluttering plumage of flags. There was one thing not lively; the officers of these gorgeous troops sat in silent dignity on campstools in front of the line—a kind of military coma that the hustling regiments now tackling the great Northern Bear in Manchuria evidently have not inherited and could not comprehend.

The situation was unique, not likely to be forgotten by any who participated in it, either American or Japanese. It was a clear and calm summer morning, as our lines disembarked and formed on the beach, the commodore stepped into his barge to follow us. Instantly the black "fireships" were wrapped in white clouds of smoke, and the thunder of their salute echoed among the hills and groves back of the village. To the startled spectators on shore they must have seemed suddenly transformed into floating volcanoes. And when the great man landed, they gazed with wonder, for no mortal eye (no Japanese mortal) had been permitted to look upon him before. In all the negotiations hitherto he had played their own game and veiled himself in mystery. They could communicate with so lofty a being only through his subordinates. This was not child's play. It was not an assumption of pomp inconsistent with republican simplicity. Commodore Perry was dealing with an oriental potentate according to oriental ideas. He showed his sagacity in doing so. At this time he was fifty-nine years old, a man of splendid physique and commanding presence. He had already lived through a varied experience that had helped to train him for this culminating achievement of his life. Endowed with strong native powers he had risen in mental capacity and executive force with every stage of his professional career.

The War of 1812, in which also his famous brother Oliver Hazard and two younger brothers served, gave him his first baptism of fire; and later the Mexican War, service in various parts of the world civilized and savage, duties on naval boards at home, investigations and experiments in naval science, naval architecture, naval education—these and numberless other methods of serving his country both in the professional routine and in general affairs, had developed his judgment, his mental acumen, his breadth of vision, his knowledge of men; and thus had prepared him for his high mission as ambassador and diplomat. Unquestionably his insight into the oriental mind, his firmness and persistence, his stalwart physical presence, his portly bearing, his dignity, his poise, his stately courtesy were prime factors in his success as a negotiator with an Eastern court. He was the right kind of man for America to send on such an errand to such a people.

On his arrival we marched to the hall through an avenue of soldiers, our escort being formed of sailors and marines from the four ships. Leaving the escort drawn up on the beach, the forty officers entered. We found ourselves within a broad canopied court of cotton hangings, carpeted with white, overlaid in the center with a scarlet breadth for a pathway leading to and extending up on the raised floor of the hall beyond. Many two-sworded officials in state robes were kneeling on either side of this flaming track. Within the hall sat— not in Japanese fashion but on chairs—the imperial commissioners, the princes Idzu and Iwami, surrounded by their kneeling suite. They were both men of some years, fifty or sixty perhaps; Idzu a pleasant intellectual-looking man, Iwami's features narrow and somewhat disfigured by the smallpox; both were attired in magnificent robes richly embroidered in silver and gold. Vacant seats opposite the commissioners were taken by the commodore and his staff. Between the lines were the interpreters, on one side a native scholar on his knees, on the other erect and dignified the official interpreter of the squadron, S. Wells Williams LL.D., a well-known author and missionary in China. Behind them stood a scarlet lacquered chest that was destined to receive the fateful missive for conveyance to court. Overhead in rich folds drooped the purple silk hangings profusely decorated with the imperial arms and the national bird, the stork.

I had scarcely noted these few details and glanced at the genial face of Bayard Taylor as he stood behind the commodore taking notes, when the ceremony began. It was very brief. A few words between the interpreters, and then, at a signal, two boys in blue entered followed then by two stalwart Negroes, probably the first to be seen on the landscape of Japan. In slow and impressive fashion the two men brought in the rosewood boxes that contained the mysterious papers. These were opened in silence and laid on the scarlet coffer. Price Iwami handed to the interpreters a formal receipt for the documents. The commodore announced that he should return the next spring for the reply. A brief conversation in answer to a question about the progress of the Taiping Rebellion in China, and the conference closed, having lasted not more than twenty minutes. A short ceremony, and witnessed by not more than fifty or sixty persons out of the entire populations of both the great countries engaged; but it was the opening of Japan. It brought together as neighbors and friends two nations that were the antipodes of each other not only in position on the globe but in almost every element of their two types of civilization.

That the Japanese have themselves appreciated the significance of this memorable meeting appears in the amazing historical developments that have followed all over the empire along the lines of commerce, industrial art, education, and religion, and is shown also by innumerable public utterances from the platform and press; and they have recently commemorated the occasion by erecting a monument at Kurihama in honor of the American commodore. But this later material can wait until the end of the chapter; we will keep on here with the main story.

This first act of the mission was now achieved, and the squadron rested from its labors. A great weight was lifted off its mind. The next day, with lightened conscience, it set itself to the easier task of surveying and sounding the bay, exploring future harbors, locating islands and rocks, measuring distances, and plotting charts. These uncanny operations were watched with some solicitude by the coast guards. They offered no active opposition, though once or twice we had occasion to show how thoroughly each boat was armed and ready for emergencies. The *Saratoga*, not willing to be outdone in this hydrographic work, located one shoal with undoubted accuracy by running upon it full tilt. Fortunately the wind was light and the bottom smooth; no harm was done to either ship or shoal. We were not proud of the achievement; but the commodore did us the honor to immortalize it and us by naming the sandbar the "Saratoga Spit"; and that title it bears to this day. Some years later it acquired a tragic interest when the U.S.S. *Oneida*, coming down the bay to sail for home, was run into in the night and sunk by the British mailship Bombay. She went down close by the "Saratoga Spit," carrying with her most of her hapless crew.

A few days after the Kurihama conference we left the Empire of the Rising Sun and returned to the Central Flowery Kingdom. On the seventeenth of July, as silently as they had entered nine days before, the two frigates steamed out of the bay with the two ships in tow. Outside they separated and went their several ways; the two steamers and the *Plymouth* back to Lew Chew and the *Saratoga* to Shanghai. We parted in a storm. If our Japanese friends could have seen our belabored ships scuttling away into the darkness and foam they would have taken it for a special interposition of their wind-god, wreaking vengeance on the Western barbarians for their temerity. The gale grew into a tempest, and the tempest into a typhoon, the largest though not the most vicious of the four encountered by the Saratoga in those uneasy seas. We compared the logbooks afterward of several ships that were caught in different sections of its enormous circuit and found that it was more than 1,000 miles in diameter and, in its progress, swept over the larger part of the north Pacific Ocean. It raged for several days, and every vessel in our fleet got entangled in some part of its vortex. Our own ship, the *Saratoga*, was under orders for Shanghai; and after the gale struck us, with battened hatches and sea-swept decks, we rode on the outer rim of that cyclone almost all the way back into the mouth of the Yang-tse-keang. It was riding a wild steed, as all sailors know who have tried it, but we got to Shanghai all the quicker. Six months we lay there at anchor off the American consulate. It was the time of the Taiping rebellion. As if to give us further object lessons in the oriental way of making history, one night the Taipings inside the walls rose and captured the city. The imperialist forces came down from Peking to retake it. And about once in three days we were treated to a Chinese battle—sometimes an assault by land, sometimes a bombardment by the fleet of forty or fifth junks; all very dramatic and spectacular, occasionally tragic, frequently funny. But, as Kipling says, that is another story and deserves a chapter of its own.

Meanwhile here is the place for a codicil in which to record the Kurihama celebrations just referred to. During autumn of 1900, Rear Admiral Beardslee, retired and traveling in Japan, took occasion to revisit the scene of the famous landing. In 1853 he was a young midshipman on board *the* Plymouth, and was in charge of one of the boats of the flotilla. He easily identified the spot and finding it neglected brought it to the attention of the Beiyu-Kwai—"Society of Friends of America"—who assumed the patriotic task of renovating the place and commemorating the event. The occasion truly was an inspiring one. On 14 July 1901, which was the forty-eighth anniversary of the conference, and on the spot where the hall of conference stood, there assembled a distinguished company of dignitaries of the empire, the officials of the Beiyu-Kwai, Admiral Beardslee and other representative Americans, together with many thousands interested spectators. Baron Kaneko presided and addressed the company. Other addresses followed, from the American minister Colonel Buck, from Viscount Katsura, from Admirals Rodgers and Beardslee, U.S.N., and also from the Governor of Kanagawa. It was felicitous circumstance that when the supreme moment came the monument was unveiled by Admiral Rodgers, a grandson of Commodore Perry and at that time commanding the American squadron in the East. The memorial is a shaft of unpolished granite standing on a massive base and rising to a height over all of thirty-three feet. The side facing the bay bears this inscription in Japanese:

Monument to Commodore Perry in Kurihama.

This monument marks the landing place of Commodore Perry of United States of North America. Marquis Ito Hirobumi, Highest Order of Merit.

On the reverse is an inscription in English:

This monument commemorates the
first arrival of Commodore Perry,
Ambassador from the United States of America,
who landed at this place July 14, 1853.
Erected July 14, 1901.

This solid memorial will forever dignify the little Japanese hamlet of Kurihama as the birthplace of the new Japan and the scene of the beginnings of a great international friendship.

In this epilogue belongs also the record of another celebration more recent and of a more personal flavor. The Japanese in foreign lands have a patriotic custom of strengthening the home ties by celebrating the birthday of their emperor, which falls on the third of November. The year 1903 marked a half century from the first landing of the Perry expedition on Japanese soil, and Mr. Uchida, the Japanese consul-general in New York, conceived the happy idea of adding still further prestige to the usual celebration by commemorating that famous event. Invitations were issued to the descendants of commodore Perry and to the now few survivors of the fleet. Out of the more than two thousand officers and men who composed the personnel of the expedition, less than a score are known to be living, three of whom were present at the reception.

To these three, who had not met for a half a century, it may well be imagined the occasion was impressive, not to say thrilling. The forty or fifty Japanese gentlemen and the ladies present, and as many more Americans, some of them descendants of the famous commodore, and others who had been resident in the Mikado's dominions or were specially interested in the country and its people, made a most brilliant assemblage. The memories were indeed inspiring.

The half century had enlarged the dimensions of the event; rather had brought out and developed its natural results along the lines of

trade, industrial art, commerce, education, intellectual and moral en-
lightenment, and so splendidly that the growing light reflected back
on the original act and revealed its magnitude. With these sentiments
were also mingled tender thoughts of shipmates long since gone and
memories of scenes that made us sigh:

> —for the touch of a vanished hand,
> And the sound of a voice that is still.

On the walls hung a large, old-time colored lithograph representing
the landing at Kurihama, a print struck off soon after the return of the
fleet and loaned for the occasion by one of the commodore's daugh-
ters. Near it was a companion picture of the same size, a photograph
of the Perry monument at Kurihama.

After the social hour, the consul called his guests to order and
made an address of welcome, alluding to the emperor, the expedition,
and the presence of some who had been members of it. Two other brief
speeches were made, one by Admiral Rodgers, a grandson of the old
commodore, who spoke of his grandfather's mission to Japan, his own
service in the East, and the unveiling of the Kurihama monument.
The other was by one of the survivors and was, or course, largely
reminiscent of those distant scenes and descriptive of our famous old
commander. It gives one a funny sensation to stand before a brilliant
company as a relic of some ancient bit of history, and be watched by
such curious eyes while you step out of your own past generation into
the light of modern times to make your speech!

When the tables were brought in for the banquet, it fell to us
three "relics" with three or four friends to surround the same board—a
sumptuous improvement on a middies' mess in the steerage of a man-
of-war and seasoned with high memories. As we broke bread together

and the current of converse moved swiftly on, it seemed almost as if we were surrounded by the unseen forms of messmates who had long since sailed on to the haven beyond. And back of all was the thought of the Sunrise Kingdom herself, the hermit land of half a century ago, so exclusive, so mysterious, but now so teeming with the activities of a new civilization, the resources of a new power, and all the dignity and responsibility of a new place in the world. The occasion itself, the sentiments it inspired, and the distinguished company uniting in the celebration, all combined to make it a memorable evening.

PART TWO
ADVENTURES BY LAND

CHAPTER 5

IN THE HOME OF THE NOMADS

by Sven Hedin

Bridge in Habba Kadal, Srinagar, 1903. Photo by Geoffroy Millias.

On a summer's night in 1906, I settled myself comfortably on the grass under the ancient plane trees of Ganderbal. The moderately warm breezes of Kashmir caressed the trunks and whispered in the crowns, but the grove was dark and the silence was broken only intermittently by nocturnal sounds, after the day had gone to rest.

Why had they not arrived yet, I pondered. Perhaps they would not appear before the new day had risen over the mountains?

"Hello," I called to the five oarsmen, who had brought me here and who were still busy with their long, slender canoe, "Light a fire so that the caravan may find us."

Dried branches crackled and cracked and tongues of flame fluttered as golden pennants in a wind. The plane trees towered in a ring of gray specters, while the crowns turned as green as the enamel in a Mohammedan mosque. The stars that had just peeped through the leafy arches were extinguished, but the grove was flooded with light as for a temple festival, and the smoke ascended like a sacrificial tribute from an incense-burner. I lit my pipe and mused. Another march of conquest was about to begin through Tibet. The long journey through Europe, the Caucasus, Asia Minor, Persia, Seistan and Baluchistan had been completed to Simla, where Viceroy Lord Minto informed me that London had refused permission to use India as a starting-point for a march into the forbidden country. Therefore I had been forced to revise my whole plan. I had gone to Srinagar, whence I would continue to Leh in Ladakh and join the main caravan route to Chinese Turkestan, detour in uninhabited regions and, unnoticed, turn eastward. The British residentiary in Srinagar had notified me that the road to Chinese territory was also closed to me, unless I had a Chinese passport. My telegraphic request had been successful, with Swedish diplomatic assistance, and the passport arrived in good time. The first caravan was assembled in Srinagar, led by Kashmirians and escorted by two well-armed Afghans and two Rajputs. Robert, a young Eurasian, was to be my secretary and the Hindoo, Manuel, my cook. We purchased thirty six fine asses from the Maharajah of Poonch, but the baggage was to be carried to Leh by hired horses. Our travel fund consisted of gold and silver rupees, current in Tibet.

The horses were loaded in my yard in Srinagar on the sixteenth of July and the long train vanished in a cloud of dust on the road to

Ganderbal. Alone I walked to the bank of a canal, where a canoe was in readiness with its oarsmen. I took my place at the rudder.

The highly polished boat glided like an eel through the water that seethed around the stern. The broad blades of the oars were bent by sinewy arms. Picturesque houses with many balconies lined the banks. Children played at landings and bridges, while women were washing linen. One house was built like a bridge across the canal, and the quaint perspectives succeeded each other so rapidly that we could not digest them before new ones were opened as we traveled on our narrow waterway. Now we were in the shade, and now the sun scorched between groves of trees and houses. Ducks and geese rooted in the slime and gnats were having their evening dance over the water.

The sun sank and twilight spread over the enchanted region. The last houses disappeared, the outlines of parks and groves suggested dark phantoms on both sides of the canal. The night was raven-black as the oarsmen slowed down and the canoe glided toward the landing at Ganderbal.

And here I was by the fire under the plane trees awaiting the caravan. There was a rustling in the bushes. By the light of the fire I recognized one of the Afghans. He whistled shrilly. Robert and Manuel also appeared, accompanied by a long row of Kashmirians.

New fires crackled. A few of the men hurried back with resinous torches in their hands to light up the trail among the trees for the missing ones.

At the midnight hour all were here. What a din, what a buzz of voices and cries! The escorts were shouting their commands, Kashmirians wrangled and quarreled, horses were neighing for their bags of corn, mules kicked and fires crackled. But, gradually, it became comparatively quiet and the white turbans were grouped around the camp fires in front of the tents. Wild faces, browned to a copper color by India's sun, glistened like metal in the glare of the fire.

Dinner was served and for the first time I entered my tent which was shared with two cute puppies, Brown and White Puppy.

At last I was ready for bed and snuffed the candle. Sleep was slow in coming. Reflections from the fires danced on the tent canvas, and the murmur of voices was audible for a long time.

A new expedition had started. In fancy I heard the roar of mighty rivers, howling of raging snowstorms and temple songs in adoration of Buddha. Endless Asia was stretching yonder, waiting for me, and mysterious Tibet with its last geographic secrets, its temple cities, Lamas and incarnated gods. My head was like the workshop of a smith, where marvelous conquests and wild adventures were being hammered out. I knew which parts of earth's highest and most expansive mountain region had remained absolutely unexplored by the Western World. The most recent maps of Tibet still showed three large white areas, in the north, in the center, and in the south, marked "unexplored." The southern area of 65,000 square miles, situated north of the Brahmaputra, was the largest, and was exceeded in size only by the polar regions and interior Arabia. I wanted to cross these unknown expanses and fill out the blank spaces on the map with mountains, rivers, and lakes, and I had an ambition to be the first white man to stand at the source of the Indus, which Alexander the Macedonian believed he had discovered 2300 years ago. I also dreamed of going through to Tashi-lunpo, the monastic citadel, where the holiest man of Tibet resides, the Tashi Lama. In the previous year he had visited India and its Viceroy, who had given me a most sympathetic account of him. By the course which I had outlined, I could not reach the Tashi Lama's holy temple-city without traversing the three white areas. In 1904 the Dalai Lama had fled to Urga and Peking, when the British army of invasion under Younghusband forced itself into Lhasa, leaving four thousand slain Tibetans along the way. The Tashi Lama was now the foremost man in Tibet. I had an almost superstitious conviction that he alone had the power to open all gates for me.

Thinking upon these matters did not induce sleep. I also wondered what would be the fate of all the men and animals that I had taken with me upon these roads of great adventures. Little could I then divine that not a man, not an animal, now stirring noisily around my tent, would be with me upon my return to Simla, two years and two months later. They were scattered as chaff before the wind. But now, as the fires died and silence enveloped the grove, all of them slept peacefully under the plane trees of Ganderbal. Our winding road led among willow, walnut, and apricot trees up through the valley of the Sind and rural villages, over swaying bridges to the music of the roaring white-foamed river. Alternately, the sun burned, or darkness followed the rapid marshaling of masses of clouds by the monsoon. We were refreshed by soft summer that filled the air. At night we listened to the mournful howls of jackals and, by day, to the tinkling bells of the caravan. The march was ever upward toward snow-capped peaks that were dyed in purple as the day dawned.

In hundreds of precipitous bends of the mountain road, we moved up the pass to Zoji-la. It became necessary to reorganize the caravan completely in the village of Kargil on the other side. The Kashmirians and Afghans, who, in true bandit fashion, had stolen sundry articles from peaceable villagers along the road, were now dismissed together with their horses. Other men were engaged and seventy-seven horses were hired for the journey to Leh.

We were now at the Indus, where we rode on narrow, breakneck paths. The Himalayas were in the rear and we were getting deeper and deeper into new labyrinths of magnificent mountains. We had left the world of Hindoos and Mohammedans and were now riding through sections where Lamaism is supreme. Here and there we passed by a picturesquely located Lamaistic monastery. Along the road long stone walls had been erected, capped with slabs of green slate, in which the sacred phrase had been inscribed: "Om mani padme hum." Lizards, as green as the slabs, darted unconcernedly over the sacred words.

The Himalayas.

I had ridden over this road twice previously, but in winter, when the ground was covered with snow. Now, the mountains were caressed by warm breezes and foaming, white, wild brooks tumbled from their sides into the Indus, to die in its embrace. I cast longing glances along the course of the mighty river and wondered if fortune would favor me by raising my tent up yonder by its source. No white man had ever been there.

Leh is one of the most charming cities of Asia, situated at no great distance from the banks of the Indus and surrounded by regal mountains. The old picturesque royal castle rises high above its stone houses, Lamaistic temples, mosques, bazaars and poplars. The body of the main caravan of invasion was to be set up and organized here, in which work I had the invaluable assistance of the Joint Commissioner, Captain Paterson, and of the lovable Moravian missionaries, who had become my friends in earlier expeditions.

But the greatest help came from Mohammed Isa, who had traveled with several European expeditions in innermost Asia and who was to be the leader of my caravan. He spoke Tibetan, enjoyed a good reputation in entire Ladakh, maintained excellent discipline, but was also good-natured and had a sense of humor. I greeted him in a friendly manner and within five minutes he was enlisted in my employ. He was given the following order:

"Engage twenty-five reliable Ladakhs, buy about sixty prime horses and provisions for at least three months."

On the following day my yard was transformed into a market and we were soon the owners of fifty-eight horses. The caravan also numbered thirty-six mules, thirty hired horses, and seven yaks, that were led by their owners.

A few days later Mohammed Isa announced that twenty-five men were on exhibition in the yard for inspection. Eight were Mussulmans, seventeen acknowledged faith in Buddha and the holy men of the Lamaistic religion. Guffaru, the oldest man in the company at sixty-two years, had brought his own funeral shroud to be assured of honorable obsequies in the event of being overtaken by death during the journey. Tsering, a brother of Mohammed Isa, also advanced in years, was to be my cook. The others will be introduced later, as they appear in their own roles. All were Ladakhs, with the exception of

Rub Das, who was a Gurkha from Nepal. All spoke Tibetan and East Turkish. During the years I had become sufficiently familiar with the latter language to express myself and could therefore use any one of my servants as interpreter.

Gulam Rasul, a rich merchant in Leh, helped us to make our purchases for the men and animals. My yard was a workshop, packsaddles were sewed for the animals, tents were made for the men, rice, flour, barley, corn, brick-tea, preserves, and numberless other articles were weighed and put in sacks, while the bells tinkled and the men talked. The sunshine filtered through the leaves of the apricot trees and cast green shadows on my floor. On an appointed day the excitement of the camp rose higher than usual. The bells on the mules gave the signal for the departure of the first division under Tsonam Tsering. Mohammed Isa followed with the main caravan.

On the fourteenth of August I started with a few men and nine baggage horses. The whole city was out to bid us farewell. Our road led out through the gate of the city by the Mohammedan burial place. Two horses shied, threw off their burdens and ran away among the markers on the graves, under which the sons of Islam await resurrection and the joys of Paradise.

After this incident all went well. Majestic mountains were rising to the left, at the right was the mighty waterway of the Indus.

The tents had already been pitched at the base of Tikse, a Lamaistic monastery, and Muhammed Isa pointed out the arrangement of the camp and the long rows of pack animals standing there, munching grain from their feedbags. Night came with rest, and silence was broken only by the songs of the sentinels.

The train proceeded deeper into Asia. On the crest of the pass, Chang-la, 17,580 feet above sea level, a cairn stood, with sacrificial sticks, covered with tattered streamers, torn by the wind. Skulls of antelopes and yaks adorned the cairn. When hailstorms beat upon the whitened foreheads, the illusion was almost complete of the whining and moaning of the dead animals.

Villages were less frequent and finally there were none. In the last ones we purchased thirty sheep, ten goats, and a pair of large, half-wild watch dogs. The two puppies from Srinagar, irritated by the first snowfall, stood in the opening of the tent and barked themselves hoarse at the falling flakes.

In the very last village the men of Ladakh celebrated a farewell feast in honor of their homeland. The entire population gathered around our campfire. Men played flutes and beat drums while the women danced.

From this point we penetrated the wilderness. On the pass, Marsimik-la, 18,340 feet above the sea, the first horse collapsed. The next pass was called Chang-lung-yogma and had an altitude of 18,960 feet. The ascent was incredibly steep and hours were needed to climb the dizzy height.

The view, that opens to the south, defeats any attempt of description by words. The valley that we had followed narrows into a mere furrow in a confusion of cliffs and ridges. The silver-white, sun-lit peaks of Himalaya tower over and above one another to the rim of the horizon. Fields of eternal snow glitter in color tones of blue, while the light-green armor of the glaciers reflects the rays of the sun in dazzling daylights of splendor.

We were on the mountain chain Karakorum. Toward the south we beheld Himalaya, in the north, Kuen-lun, the border wall to Chinese Turkestan. Desolate Tibet expanded toward the east and southeast.

This wonderful scenery was quickly blotted out by a chilling snowstorm, while the long dark line of the caravan proceeded to the Tibetan Highland. We encamped on a spot as desolate as the surface of the moon, twenty-five hundred feet higher than the peak of Mt. Blanc. Not a blade of grass grows here. The rainfall in the area does not reach the Indus but runs into small basins without any outlets. Geographical names are totally wanting and I shall designate our camping places by numerals.

Only a few days before we had enjoyed summer. We had now been received by the most inhospitable winter. We were wrapped in darkness in the middle of the day while hail pelted us and finally turned to snow. We marched in four columns quite close to each other. The animals with their burdens and the men on their horses were chalky white—we resembled sculptures in alabaster. We were able to see our nearest neighbor only, and simply followed the tinkling of the closest bell.

All changed in a few days. The weather cleared to radiance and the ground dried. We even suffered from the lack of water, but later discovered a spring at the foot of a mountain to the northeast, glistening like silver in the sunlight.

Our chosen course then led eastward in a valley, twenty miles wide, where antelopes and wild asses had undisturbed grazing and where our own animals found nourishment. We were not yet in an unknown land.

We camped several days on the western shore of a large lake that was discovered by Captain Wellby in 1896. While here we dismissed the owners of the hired horses and yaks, as well as the men from India, who could not stand the severe climate. And now the final tie that bound me to the outer world was severed. The returning men carried my last mail back with them.

The large lake is oblong, running east and west. Kuen-lun rises in the north, while in the south is a chain of wild, precipitous cliffs, changing in red and flaming yellow colors with the same wild intensity as in the Grand Canyon at sunset. The peaks are shaped like pyramids and cupolas with shining caps of eternal snow, and in the valleys between the mountains, blue and green glaciers extend their armors of ice toward the lake. The sky is turquoise blue; not even the slightest breeze ruffles the lake, whose smooth surface reflects the fantastic contours and brilliant colors of the mountains.

We moved our camp to the north shore. While Mohammed Isa was conducting the caravan to the east end of the lake, we put our boat in position. I sat at the helm and Rehim Ali was my oarsman. The distance to the south shore looked to be short. I should have time to make a series of soundings and reach the camp at the eastern shore before nightfall. A signal fire was to be kept burning in the camp if we were delayed.

We started off. The depth was one hundred and sixty feet. A little later the sounding-lead of the line, two hundred and thirteen feet long, did not touch bottom.

A deathly silence surrounded us, broken only by the splash of the oars and the ripples around the stern. The smooth mirror-like sheet was cut by the boat. We were gliding along in a landscape of dreams. It was perplexingly difficult to determine where the fiery-red mountains ended and the reflection began. The mirrored image of the heavens at nadir was just as exquisitely blue as in zenith. One became dizzy and had the sensation of soaring through crystal-clear space within a ring of glowing volcanoes. Finally we reached the desolate shore. It was late in the afternoon. We again put out and steered toward the east. An hour passed; Rehim Ali looked uneasy. Upon my question about the reason, he answered: "Storm."

I turned around. The horizon in the west darkened and yellowish-gray dust clouds swept over the mountains. A roar was heard in the distance. The lake was still as smooth as glass. But the heralds of the storm were over us.

"Hoist the mast and sail." The boat was rigged in a moment. I grasped the sheet and the tiller. At the first gust of wind the sail filled and our light craft shot like a frightened duck over waves that soon grew to billows. Swiftly as an arrow we glided by the flat sandy points that jutted out from the shore. A flock of wild geese sat on one of the points amazed at the big bird that used only one motionless wing.

The next point extended far out into the lake, encircled by seething breakers. If we failed to clear it, we would be shipwrecked, for the oiled canvas was stretched like a drumhead over the wooden braces and would be rent in colliding with the bottom in this mad speed. The storm raged in all its fury. The mast was bent like a bow. Foam-crested waves raced by us and the water in our wake seethed in millions of boiling bubbles.

The atmosphere had cleared. The sinking sun resembled a ball of glittering gold. Scarlet skies were driving eastward. As if illuminated from within, the mountains glowed like rubies. The storm atomized the foam on the waves and scarlet plumes floated like flying veils over the lake. Shadows were lengthened, only the highest peaks still being gilded by the setting sun.

Over the fore-top the white breakers were seen around a new point. We must veer to starboard and land alee to await dawn. But the maneuver was impossible. The sea was too high and the wind too strong. In a few seconds the booming around the point was inaudible, for we were being driven out on endless wastes of water, over which the wings of night were spread.

The moon rose over the mountains to give a silvery touch to the foamy wreaths on the crests of the waves, which were chasing one another like threatening specters. We were flying directly east. The life belts were ready, for if the boat were filled it would immediately be drawn down by the center board.

All my strength was needed to prevent the boat from steering against the wind. I looked in vain for the beacon fire. The moon set. The darkness was impenetrable, the stars alone twinkled. It was killing cold. The spray from the crests of the waves turned into an armor of ice on our garments. The whole night was ahead of us; in the east an unknown shore, where we might be hurled against perpendicular rocks and crushed in the darkness and breakers.

A dull roar was heard over the foretop. It was from the breakers on the beach. We were hurled ashore by the roll and suction of the waves. Everything was saved. We were soaking wet. It was –16 degrees Celsius. We tilted the boat against an oar and had shelter. We kindled a small fire with difficulty. My feet had become numb and Rehim Ali rubbed them.

We were hungry, tired, and nearly frozen, when we heard the hoof-beats of horses. Mohammed Isa and two men had come to our rescue. They had believed us drowned in the storm and had just started out to look for us.

A few days later we camped again on the west shore of a bitterly cold salt lake, which also had to be sounded. Supplied with provisions, sweetwater, and warm clothing, Robert, Rehim Ali, and I rowed across to the north shore in glorious weather and had our breakfast.

The greatest depth was only fifty-two feet. The shores were low and the bottom consisted of a deposit mixed with salt in hard, keen-edged cakes. A row of these blocks of salt extended out into the lake at our landing place. We walked on them and drew our boat into somewhat deeper water.

We stepped aboard. At that moment the western sky took on a threatening aspect. We raised the mast and the first gust of wind caused the sail to flutter.

"Perhaps it is safer to spend the night here." We had provisions and fur coats and would have time to gather yak dung before it was dark.

As we were about to land we saw two large, light gray wolves standing on the shore at a distance of fifty feet with dripping tongues and eyes aglow with hunter and blood thirstiness. Neither the fluttering soil, nor the rocks of salt that we threw, frightened them away. They seemed to understand that we had not brought weapons to a lake of saltwater, where no fowl are found. I had previous proofs of the wolf's unlimited audacity and if these two specimens were the heralds of a whole pack we might have an uncomfortable night on the shore.

The wolves paced back and forth impatiently. The storm increased and the waves rose high and white. Our choice was between the wolves and the storm. A sudden squall moved the boat from the shore. My oarsmen took their places. The wolves contemplated us with disappointment and anger and trotted eastward along the shore, sensing that sooner or later we must land.

With wind and waves from starboard we crossed the lake and fought for our lives. The greatest peril was landing among the sharp salt-slabs where the boat would be split like paper. As the depth was only six feet, I intended to turn and take a chance on the open water among rolling, silvery waves, rather than suffer shipwreck on the shore.

In that moment the undulations ceased and we discovered that we were in calm water. A salt point that we had not noticed gave us shelter. That night, spent on this wet shore of salt in a biting cold, I prefer to pass over in silence. It was beastly, and endlessly long. On the following morning we found the camp, where we had hot tea, wrapped ourselves in furs and slept like the dead for a whole day.

We had now come to the stage where hardly a day elapsed that we did not lost one of our animals. Eighteen horses and two mules had died. A pack of wolves followed us faithfully and gorged itself on the fallen martyrs. A death watch of six ravens had followed us for six weeks. The black birds of death laughed hoarsely at the attempts of the puppies to drive them away. When a storm approached they sounded cries of alarm. As soon as a horse had died, they pecked his eyes out. Nevertheless, we would have missed them, if they had deserted us.

The herd of sheep was soon gone. Tundup Sonam, the hunter, provided us with daily meat. He sometimes killed a yak with one ball straight to the heart. Once he killed two wild sheep. Quite often he came lugging an antelope, whose meat was better than that of other animals.

We halted for a day or two at fairly good pastures. Our pack animals were not tethered at night, but were guarded on account of wolves. The mules were more sensitive to the cold than the horses. On the night of October seventh, when the temperature sank to −24 degrees Celsius, a few mules stationed themselves at the door of my tent. They knew that the tent gave shelter against the cold.

We once missed three horses. Robsang started out afoot and alone to look for them. He was absent three days. I feared that the wolves had devoured him, but in the evening he returned with two horses. The third horse had been driven by the wolves to the shore of a small salt lake, that was still open, where they expected he would turn back to become the victim of their fangs. The horse, however, as his tracks indicated, had jumped into the lake to swim to the other shore, but strength failed him and he was drowned. Robsang believed that the tracks of the wolves betrayed their confusion and disappointment.

Mohammed Isa started off with the horses in the direction I had indicated. After breakfast I rode on with Robert and Rehim Ali, who held my horse at all stops. When we came by dried yak dung, we built a fire to warm my hands sufficiently to take notes and sketches.

Tsering was always the last person to leave. He was responsible for my tent and for the baggage that I needed daily. His little caravan usually passed me at one of the resting places. Once I caught up with him on a pass, where he exerted himself to build a cairn.

"What good will that do? We are the last ones in the train."

"To appease the mountain spirit and give us a safe pilgrimage to Tashi-lumpo," he answered.

All the Lamaists were just as desirous as I that the pilgrimage would be made successfully.

Winter was now setting in. On October seventeenth the cold was −2 degrees Celsius. At that time I had an equal number of men, horses and mules, or twenty-seven of each kind. Superfluous baggage was discarded. I gave up several books. In two months we had not seen the trace of a human being.

Tsering sat by the fire preparing my dinner in a violent snow-storm. In the meanwhile he told stories to the others. Snowflakes sputtered in the fire.

By October twentieth the whole country was white. The cara-van proceeded to a pass. I trailed, as usual. The snow became deeper.

There, a horse had fallen! The ravens had already pecked his eyes out. The wind had driven the snow up against his back. He lay as though resting on a bed of white sheets.

The pass was 18,400 feet above sea level. Icy winds. Ten degrees cold, impenetrable snowstorm. Confused by the snowstorm, Mohammed Isa had chosen the wrong course. We must remain together. It would be perilous to lose each other after all tracks had been obliterated. I followed his tracks in the snow and we camped in an abyss, almost snowed in. We hoped for Nomads who could sell us yaks and horses.

In the morning one mule was dead at the camp and two horses had collapsed nearby. At Camp Number 46 there was no grazing. The horses chewed each other's tails and packsaddles. The latter were stuffed with hay. Forty days' rations of rice were reserved for the men, the balance, as well as barley and corn, was fed to the pack-animals.

Mohammed reconnoitered the nasty labyrinth of snow-covered mountains into which we had been misled. He had discovered level ground with pasture in the southeast. At twilight he asked for permission to conduct the caravan there. I remained with Tsering and three other men.

The cold sank to −27 degrees Celsius. In the morning one mule was dead—frozen stiff. If we had raised him, he would have stood like a horse in a gymnasium. Another one died as the sun rose and the reflection of the rays gave almost a sign of life to the open eye.

I started out with the shattered wreck. We met Tundup Sonam, who had been sent to show us the road. The caravan had lost its way in the night and became divided. Four mules had died; herder and sheep had disappeared. The caravan was facing destruction. If we could not find Nomads soon, we would be compelled to throw the baggage away and continue on foot. We were piloted by the hunter in biting cold and a blinding snowstorm. In time we reached the plain, built a fire so that we would not freeze to death. A little later we met a scout, who directed us to the division that was led by Sonam Tsering. Nine pack animals had died in this terrible night. Twenty-one emaciated horses and twenty mules in the same pitiful condition remained. We could not proceed far under such conditions. Four hundred miles separated us from Dangra-yun-tso, the lake, to which I had asked the private secretary of the Viceroy to dispatch my mail by special courier. The goals I had set seemed to be out of reach.

The other part of the divided caravan was reassembled. The herder had tethered the sheep in a ravine and sat among them to keep from freezing to death. By a miracle they had escaped the wolves.

Wooden boxes were burned in the evening fires, dispensable cooking utensils, felt carpets, and horseshoes were discarded. The other articles were put in sacks. Tundup Sonam shot three antelopes, one of which had been devoured by wolves before we could get him.

A short day's march led us to a small, deeply frozen lake. Late in the evening a flock of wild geese flew on its way to warmer regions. Their chorus of honks indicated that they had intended to settle by a spring at the shore. When they saw their rendezvous occupied, they rose higher and their honks died away in the distance. Their forebearers had traveled the same course in falls and springs. They journeyed in snowstorms and sunshine, night and day. By light of moon and stars they saw the little lake gleam like a shield of silver. I envied them. In a few days and nights they traversed the entire Tibet and lifted themselves over the highest mountain chains on earth, while we needed months and our animals were perishing.

Wherever Tibet is crossed, wild geese may be seen every spring and fall. Separate colonies travel on different, traditional, direct routes. Do they select the time of the full moon for their flight, when the earth is illumined? Tibetans have a touching reverence for the wild geese, not the least because these winged Nomads of the air practice monogamy. A Tibetan would rather die of starvation than do violence to a wild goose.

The wolves became bolder and howled just outside of our tents. The night-guard was increased. One night Tundup lay in ambush and shot a wolf that limped out on the ice and lay down to die.

On another occasion Tundup shot a yak that had two Tibetan balls in its body. He surprised a herd of yaks in a glen. They all fled with the exception of a bull who stopped and was wounded. Foaming with rage, the animal charged. In the last moment the hunter swung himself up on a terrace from which he aimed straight to the heart and the yak fell. We erected out tents on this spot and had meat for several days.

On a day when my usual companions and I were about two hundred meters away from the camp, Mohammed Isa fired a bullet into a herd of yaks that was grazing nearby. Instead of venting his fury on the marksman, a stately bull charged with full speed directly at Robert, Rehim Ali, and me. Foaming with rage, the yak lowered his horns to raise me and my horse in the air. I removed my fur coat to throw it over the yak's head in the last second. But Rehim Ali had stumbled and fallen and the yak selected him as victim, rushed over him, and continued his flight. I rode back to the prostrate man in the belief that he had been gored to death. He had escaped with a bloody streak on one leg and torn clothing. As a result of shock to his nervous system from the terrifying experience, Rehim Ali showed signs of an unbalanced mind for a time.

*   *   *

At last we were getting nearer to human beings. We noticed several old fireplaces and a placer mining camp, where gold prospectors had occupied a score of tents. One night our last horses and mules were chased northward by a pack of hungry wolves, but were overtaken and saved in time.

Tundup returned from a hunting expedition on November tenth and related that he had discovered a black tent in a valley towards the west. Upon closer investigation we learned that the tent was inhabited by a woman, who had two husbands, both of whom were away hunting. They had neither yaks, horses, nor provisions for sale.

Three horses died and only thirteen remained. A report was made to me on November twelfth: "Sir, Tundup Sonam is returning from the valley with two Tibetans." Seeing our tents they were frightened and wanted to flee, but Tundup reassured them. They laid down their guns at a proper distance and followed Mohammed Isa with lagging steps to his tent, where they were offered tea and tobacco.

Afterward they came to my tent, fell on their knees and so had come to the camp and improved our prospects at one stroke. They carried all kinds of articles inside of their baggy coats, dried chunks of meat, wooden bowls for tsamba and tea. Tobacco pouch, pipe, steel, bodkins, and knives dangled from belts in rhythm with their steps. Their black coarse hair hung in tufts around their greasy coat collars and the lice that inhabited these primeval forests had never in their lives run the risk of being caught in a comb. Musketoons and bifurcated props were strapped over their shoulders, their belts held broad swords and the men rode small, chubby, long-haired horses with bright, lively eyes.

They call themselves Changpas or Northmen and spend their winters in the desolate regions of Chang-tang in northern Tibet to eke out a living by hunting. Cows furnish them milk, butter, cheese, and cream. From big game they secure meat, skins, and fur. They prefer the meat when it is raw, hard, dry, and old. They may often be seen taking out the rib of a yak or a wild ass, more like a blackened stick of wood, from the ample folds of their fur coats, and carve it with their sharp case knives. Chinese brick-tea is the chief delicacy among the good things of life, especially if it has an abundance of leaves and stems. A lump of butter swims the wooden bowl like oil among driftwood. In this count even the horses eat meat on account of the scarcity of pasture. It seems strange to see small, grass-eating animals stand and munch the strips of meat, until the saliva hangs in long icicles from their mouths.

Not even a person of refined taste need turn up his nose at the menu placed before him in the tent of a yak hunter: goat's milk with fat, yellow cream; yak kidneys browned fat; yak marrow, toasted over dung-fire; small, fat pieces of the tender meat along the spine of the antelope, or its head held by the long horns in the flames until the skin has to be burned away and it all looks like a mass of soot; table salt found in inexhaustible quantities on the shores of lakes.

Nomads and hunters, in common with wild geese, are migratory and know all springs and pastures. They rest and hunt where their forebearers have tented and hunted. They set their traps for antelopes or lie in ambush for wild asses at a stone wall that perhaps has stood since time immemorial. Quietly and stealthily the hunter steals up on the yak against the wind and knows by experience just when to stop and shoot. Then he strikes sparks with the steel against the flint. The burning tinder lights the end of a cord, that is brought in contact with the touch-hole by the hammer, after the bead has been drawn. He does not shoot until he is sure of hitting the mark, for he must save powder and lead. The yak falls, the meat is cut up and preserved under the folds of the tent. The pelts of yaks, wild asses, antelopes, and wild sheep are tanned and utilized. Boots, harness, straps, and many other articles are made from them and the sinews serve as thread.

When the Changpa men are on the hunt, the women care for the domestic animals, and as the hunter returns at sunset the ruminating yaks are lying down in front of the tent. There they lie all night and the Nomads need not go far for dung, their only fuel. The sheep are herded into a circular fold of stone, and wolves are kept at a distance by large wild dogs.

As darkness falls the family is seated around the fire over which the teakettle is boiling. The long pipe moves from mouth to mouth. The conversation touches upon the success of the hunt, the care of the herds, and the removal to better pasture. Worn-out soles are repaired and hides are tanned by hand. A woman churns butter in a wooden stoup, while her small naked children are playing in the light of the fire. Each one retires to his own lair of furs, and before the rising sun has gilded the mountain tops the bellows are blowing to revive the fire.

Thus they live and thus they roam and thus it has been generation after generation, for uncounted centuries. Chang-tang is their poor homeland, where they live bravely and largely in God's free air, struggling against poverty and dangers. They have no fear of the roar of the storm: the clouds are their brothers. They share dominion over mountains and valleys only with the beasts of the wilderness, and the eternal stars twinkle over their tents by night. They love the icy cold, the dancing drift-snow and the white moonlight in quiet Tibetan wintry nights.

During his whole life the Nomad has the deepest reverence for the spirits of the mountains, lakes and springs. He is convinced that the hunt will be unsuccessful if he does not read devoutly his "Om mani padme hum." He knows that the spirits of the air, unless due veneration be accorded them, will bury all pastures under heaps of snow—the doom of inevitable starvation of sheep and goats—and he may well fear an unfortunate end to his wandering if he does not add a new stone to the old cairn on the mountain passes as he goes by. He does not have the slightest idea of the splendor of temple halls and of the blue smoke that circles up to the faces of the gilded gods. A pilgrimage to the great monastic cities is the privilege of rich Nomads. He believes in transmigration and is convinced that all evil deeds will be punished in the next existence, when he reappears in the form of a pack animal, a dog, or a vulture.

Some day death will stand at the tent door and peer through. The storm is howling outside, fire burns in the hearth, and silence is broken only by the everlasting prayer "Om mani padme hum." The dying man reviews his long, laborious, joyless life. He is afraid of the evil spirits, who are waiting for his departure to lead the soul on its dismal wandering into the great unknown. Unless he has appeased them while living, it is now too late. He resigns himself hopelessly to their power and caprice. Bent, wrinkled, and gray, the old hunter finishes his course. The hunting ground, where he has lived his days, vanishes back of him and he takes the first step out into the uncertain darkness. His nearest relatives carry the body to a mountain where it is laid, naked and frozen, as food for wolves and birds of prey. In life he had no continuing city and no grave after death. His grandchildren do not know where he was laid. Perhaps that is best, for where dead men's bones whiten, evil spirits dwell.

When we resumed our march, we took the two Nomads along as pilots and I could begin to insert names on my map again. They told us everything they knew about the region, roads, and the roaming of their friends. Four days later they informed us that their knowledge of the country was exhausted. I paid them four rupees a day each for the time they had been with us. Kashmir gave each one of them a case knife and an armful of empty cigarette tins. Our generosity amazed them and they declared it beyond belief that such kind people existed.

It was severely cold and the temperature went down to −3 degrees Celsius at night. Our solitary travel would soon be ended. Beyond the threshold of a pass we noticed large herds of sheep, yaks, and six black tents. A pack of half-wild dogs barked themselves hoarse and the inhabitants were astonished when we raised our tents near by on the shore of lake Dungtsa-tso.

Lobsang Tsering, the beardless and whether-beaten chief of the tent village, snuffed, laughed, and chattered. Evidently he had received no orders from Lhasa. We purchased five yaks of him and once more our twenty-five veterans were given assistance. The useful yaks had come to us a veritable Godsend.

Alternately, I rode the dapper gray horse that I mounted at Leh, or a small, lively white animal from Ladakh. The winter storms had begun. We stiffened in the saddles. Our eyes watered and the tears congealed. Garments became gray from the flying dust. Lips cracked, especially if we laughed but there were few appeals to risibilities in a temperature of −33 degrees Celsius. We wanted to get into our tents and to a fire. The storm whizzed and howled.

Four mules died. Wolves were on the spot almost before we left it.

Mohammed Isa attempted to buy yaks in a tent village. A man approached and roared in an authoritative voice.

"A European is among you. We will sell you no yaks. Turn back, or it will fare you ill."

In the night of December first the temperature was −31 degrees Celsius. One mule lay dead between the tents and had nearly been devoured by wolves before the place was out of our sight. Camp Number 77 was located in a valley among wild cliffs. Two men in many-colored furs wearing ivory tings and sacred silver caskets around their necks appeared. They carried guns and swords in silver scabbards, studded with turquoises and corals. They were members of a group of thirty-five pilgrims who, with their herds of one hundred yaks and six hundred sheep, had been at the sacred mountain (Kailas) and the sacred lake (Manasarowar). They had seen me, my Cossacks, and Lama five years ago, when the Governor of Naktsong compelled me to change course to Ladakh.

Early in the dark, rough morning I was awakened by the rattling of guns and swords as they came to sell yaks. They gave us the following information: "Orders to stop you are being circulated south of the next pass."

Consequently, we were to face the same opposition as formerly! Nearly all of our baggage was borne by eighteen yaks. A new world spread itself in the south from a pass, but it was soon enveloped in a snowstorm. We plodded on through the snow. Three men rode toward us on snorting horses, put a few questions to us, and went over to Mohammed Isa.

The summons were therefore in full action! While we were encamped a few days later on the shore of Bogtsang-tsangpo, a group of Tibetans came to my tent. Their "Gova," or chief, recognized me from 1901 and must make an immediate report to the Governor of Naktsong. He pleaded in vain with us to remain, and as we continued our journey, he accompanied us down the river for five days. We caught excellent fishes through the wakes. The Tibetans believe that lizards and snakes are equally suitable food. Karma Tamding in Tang-yung, also recognized me from my previous visit, five years ago. At that time the whole country had talked about my journey. All attempts to preserve my incognito were futile. He sold three yaks to us. Another mule died just as the first stars became visible. We now had only two mules and eleven horses. All the veterans were relieved of burden-bearing.

Karma Tamding returned, accompanied by twelve Nomads, with a large quantity of provisions. We bought toasted meal and corn for sixty-eight rupees. Two women were with him, well clothed as a protection against the paralyzing wind. We did not see much of them, but the little we saw was very dirty.

We shortened the days' marches in order to pitch camp before we were thoroughly frozen. On the morning before Christmas an old mendicant: Lama sat outside of my tent singing and swinging his magic wand which was literally covered with colored pieces of cloth, tassels, and gimcracks. He had wandered all over Tibet, begged from tent to tent, danced with his magic wand, and sung his incantations for food.

Our Christmas camp was raised on the shore of a small lake, Dumbok-tso. For dinner Tsering offered us a pan of superb sour milk, juicy mutton roasted over the coals, fresh wheat bread, and tea. I fancied hearing church bells ringing far away and the jingle of sleigh-bells in the Swedish forests. The tent was illuminated, my Ladakhs sang, and the Tibetans must have thought we were performing sacrificial rites and singing to unknown gods. Before the last candle had been extinguished I read the Bible texts for the day. The stars of Orion sparkled with incomparable brilliance out yonder in the night. We pitched our camp on the north shore of Lake Ngang-tse-tso, altitude, 15,640 feet. We decided to rest here for a season. I wanted to chart and sound the lake. From this point southward to Tsang-po, the country was unknown. Our tired animals also needed rest. The only anxiety was caused by a possibility that the alert watchmen in Lhasa should anticipate us and prevent us from continuing. We had eight horses and one mule left, and the twenty-one yaks were beginning to be footsore. We simply must give the animals a rest.

I spent nine days on the ice of the lake and sounded through the wakes. The greatest depth was only thirty-three feet. Two Ladakhs pulled me along the ice for sixty-six miles on an improved sled. Seven Ladakhs carried provisions. Our camps were pitched on the shore.

The ice of the lake was covered with fine powdered salt. Wrapped in a sheepskin coat, I sat on the sled. We moved rapidly as my two Ladakhs ran from wake to wake. One day a snowstorm raged straight in our faces. The two men were blown over and the sled was swept away by the storm and raced on until it was upset in a crack. That runaway ride was glorious while it lasted, for the wind was not felt. But, after the tumble. I faced it again. We resumed our journey. I could scarcely see the two men ten feet ahead. The powdered salt swept over the smooth ice and gave the illusion that we were moving at a dizzy speed.

We made much better time later with the wind. We rested a day in a ravine on the shore. A messenger from Mohammed Isa arrived, half dead of fatigue. He had been seeking us on the ice for fifty-four hours.

He reported that, on January first, six armed horsemen had arrived at our stationery camp and made the usual investigation. They had returned on the following day with reinforcements and a message from the Governor of Naktsong to remain where we were and that I must go back to the camp and personally answer the Governor's questions so that he could dispatch a report to Lhasa.

Had we not endured enough? Had we not lost almost our entire caravan? I imagined hearing once more the creaking of the copper gates as they closed, shutting us out from the land of sacred books, the forbidden land.

After the messenger had rested, I sent him back with the declaration that if the chief of the patrol wished to talk to me, he and his entire cavalcade would be welcome to do so on the ice.

On the following day stones and stems were white with hoarfrost. The smooth ice had become wavy as watered silk from the powder.

We sounded in new wakes. Mohammed Isa and two men found us on January sixth. My competent leader of the caravan reported:

"Sir, we had intended moving the camp today to the shore so as to be nearer to you, when three Tibetans, who were encamped close by, appeared and compelled us to unload the caravan which was ready to start, and prohibited us from taking one step southward. The Governor is expected within three days. Mounted couriers are in constant motion between him and the patrol. They repeatedly ask why you are out on the ice and seem to think that it is of no importance how deep Lake Ngangtse-tso may be. They suspect that you are getting gold from the bottom through the wakes and have just sent patrols along the shores."

Mohammed Isa returned and we continued our soundings and camped on the shore, which was already occupied by a herd of wild asses and a wolf.

Early the following morning a courier was sent to the stationary camp for my mount. When I arrived at the tent city the Tibetans sat in the doors of their tents, looking out like so many marmots in their burrows. The chiefs were eventually informed that I would receive them in Mohammed Isa's tent. They came, humbly saluting and with congress hanging out of their mouths. The leader, dressed in a red band around his head, dark blue fur coat, and with a sword in his belt, had been a member of Hlaje Tsering's suite in 1901, when we tented together on the shore of Chargut-tso.

A long conversation ensued and the chieftains confirmed that it was my old friend Hlaje Tsering, who personally would be here in a few days to pass judgment upon me and my caravan.

"Will he be accompanied by five hundred horsemen, as upon the former occasion?" I queried.

"No, Bombo Chimbo, he noticed that troops of horsemen did not frighten you. He now hopes that you will comply with his wishes."

"I have neither time nor disposition to remain here and wait for Hlaje Tsering," I answered.

"Bombo Chimbo, if the Governor does not arrive within three days, you may cut our throats."

Towards evening on the eleventh of January a body of horsemen was outlined on the hills in the east and new tents were erected around us. One of them was more ornamental than the others, made of white and blue canvas. A new troop of horsemen came shortly thereafter. The foremost man was old, bent and was wrapped in fluffy expensive furs and wore a red fur-lined bashlik on his head. After dismounting they laid their guns on the ground and crawled into the tents.

The old man was really Hlaje Tsering! I realized the weakness of my present position. I knew how hopeless it was to persuade a Tibetan Governor, either by a friendly attitude, or by intimidation, to open the roads to the sacred cities.

Bitter regret tortured me because I had not followed the original plan of proceeding to Dangra-yum-tso! It grieved me to contemplate that the large white space on the map began immediately south of Ngangtse-tso and that I must turn back from its very threshold.

True, we had traversed the unknown land to the north, discovered several lakes and mountains, sounded and charted Ngangtse-tso. But all these things were insignificant in comparison with the objective of this expedition, the exploration of the unknown land to the south and discovery of the source of the Indus.

CHAPTER 6

EXCERPT FROM
THE TRAVELS OF MARCO POLO

"The Travels of Marco Polo." Artist unknown.

Know reader, that the time when Baldwin II was Emperor of Constantinople, where a magistrate representing the Doge of Venice then resided, and in the year of our Lord 1250, Nicolo Polo, the father of the said Marco, and Maffeo, the brother of Nicolo, respectable and well-informed men, embarked in a ship of their own, with a rich and varied cargo of goods, and reached Constantinople in safety. After mature deliberation on the subject of their proceedings, it was determined, as the measure most likely to improve their trading capital, that they should prosecute their voyage into the Euxine or Black Sea. With this view they made purchases of many fine and costly jewels, and taking their departure from Constantinople navigated that sea to a port name Soldaia, from whence they travelled on horseback many days until they reached the court of a powerful chief of the Western Tartars, named Barka VI, who dwelt in the cities of Bolgara and Assara, and had the reputation of being one of the most liberal and civilized princes hitherto known amongst the tribes of Tartary. He expressed much satisfaction at the arrival of these travelers, and received them with marks of distinction. In return for which courtesy, when they had laid before him the jewels they brought with them, and perceived that their beauty pleased him, they presented them for his acceptance. The liberality of this conduct on the part of the two brothers struck him with admiration; and being unwilling that they should surpass him in generosity, he not only directed double the value of the jewels to be paid to them, but made them in addition several other rich presents.

The brothers having resided a year in the dominions of this prince, they became desirous of revisiting their native country, but were impeded by the sudden breaking out of a war between him and another chief, named Alaù, who ruled over the Eastern Tartars. In the fierce and very bloody battle that ensued between their respective armies, Alaù was victorious, in consequence of which, the roads being rendered unsafe for travellers, the brothers could not attempt to return by the way they came; and it was recommended to them as the only practicable mode of reaching Constantinople, to proceed in an easterly direction, by an unfrequented route, so as to skirt the limits of Barka's territories. Accordingly they made their way to a town named Oukaka, situated on the confines of the kingdom of the Western Tartars. Leaving that place, and advancing still further, they crossed the Tigris, one of the four rivers of Paradise, and came to a desert, the

extent of which was seventeen days' journey, wherein they found neither town, castle, nor any substantial building, but only Tartars with
their herds, dwelling in tents on the plain. Having passed this tract,
they arrived at length at a well-built city called Bokhara, a province of
that name belonging to the dominions of Persia, and the noblest city
of that kingdom, but governed by a prince whose name was Barak.
Here, from inability to proceed further, they remained three years.

It happened while these brothers were in Bokhara, that a person
of consequence and gifted with eminent talents made his appearance
there. He was proceeding as ambassador from Alaù before-mentioned,
to the Grand Khan, supreme chief of all the Tartars named Kublai,
whose resident was at the extremity of the continent, in a direction between northeast and east. Not having ever before had an opportunity,
although he wished it, of seeing any natives of Italy, he was gratified
in a high degree at meeting and conversing with these brothers, who
had now become proficient in the Tartar language; and after associ-
ating with them for several days, and finding their manners agreeable
to him, he proposed to them that they should accompany him to the
presence of the Great Khan, who would be pleased by their appearance at his court, which had not hitherto been visited by any person
from their country; adding assurances that they would be honourably
received, and recompensed with many gifts. Convinced as they were
that their endeavours to return homeward would expose them to the
most imminent risks, they agreed to this proposal, and recommending
themselves to the protection of the Almighty, they set out on their
journey in the suite of the ambassador, attended by several Christian
servants whom they had brought with them from Venice. The course
they took at first was between the northeast and north, and an entire
year was consumed before they were enabled to reach the imperial
residence, in consequence of the extraordinary delays occasioned by

the snows and the swelling of the rivers, which obliged them to halt until the former had melted and the floods had subsided. Many things worthy of admiration were observed by them in the progress of their journey, but which are here omitted, as they will be described by Marco Polo, in the sequel of the book.

Being introduced to the presence of the Grand Khan, Kublai, the travelers were received by him with the condescension and affability that belonged to his character, and as they were the first Latins who had made their appearance in that country, they were entertained with feasts and honoured with other marks of distinction. Entering graciously into conversation with them, he made earnest inquiries on the subject of the western parts of the world, of the emperor of the Romans, and of other Christian kings and princes. He wished to be informed of their relative consequence, the extent of their possessions, the manner in which justice was administered in their several kingdoms and principalities, how they conducted themselves in warfare, and above all he questioned them particularly respecting the Pope, the affairs of the church, and the religious worship and doctrine of the Christians. Being well instructed and discreet men, they gave appropriate answers upon all these points, and as they were perfectly acquainted with the Tartar (Moghul) language, they expressed themselves always in becoming terms; insomuch that the Grand Khan, holding them in high estimation, frequently commanded their attendance.

When he had obtained all the information that the two brothers communicated with so much good sense, he expressed himself well satisfied, and having formed in his mind the design of employing them as his ambassadors to the Pope, after consulting with his ministers on the subject, he proposed to them, with many kind entreaties, that they should accompany one of his officers, named Khogatal, on a mission to the see of Rome. His object, he told them, was to make a request to his holiness that he would send to him a hundred men of learning, thoroughly acquainted with the principles of the Christian religion, as well as with the seven arts, and qualified to prove to the learned of his dominions, by just and fair argument, that the faith professed by Christians is superior to, and founded upon more evident truth than, any other; that the gods of the Tartars and the idols worshipped in their houses were only evil spirits, and that they and the people of the East in general were under an error in reverencing them as divinities. He moreover signified his pleasure that upon their return they should bring with them, from Jerusalem, some of the holy oil from the lamp which is kept burning over the sepulchre of our Lord Jesus Christ, whom he professed to hold in veneration and to consider as the true God. Having heard these commands addressed to them by the Grand Khan, they humbly prostrated themselves before him, declaring their willingness and instant readiness to perform, to the utmost of their ability, whatever might be the royal will. Upon which he caused letters, in the Tartarian language, to be written in his name to the Pope of Rome; and these he delivered into their hands. He likewise gave orders that they should be furnished with a golden tablet displaying the imperial cipher, according to the usage established by his majesty; in virtue of which the person bearing it, together with his whole suite, are safely conveyed and escorted from station to station by the governors of all places with the imperial dominions, and are entitled, during the time of their residing in any city, castle, town, or village, to a supply of provisions and everything necessary for their accommodation.

Being thus honourably commissioned they took their leave of the Grand Khan, and set out on their journey, but had not proceeded more than twenty days when the officer, named Khogatal, their companion, fell dangerously ill, in the city named Alau. In this dilemma it was determined, upon consulting all who were present, and with the approbation of the man himself, that they should leave him behind. In the prosecution of their journey they derived essential benefit from being provided with the royal tablet, which procured them attention in every place through which they passed. Their expenses were defrayed, and escorts were furnished. But notwithstanding these advantages, so great were the natural difficulties they had to encounter, from the extreme cold, the snow, the ice, and the flooding of the rivers, that their progress was unavoidably tedious, and three years elapsed before they were enabled to reach a seaport town in the lesser Armenia, named Laiassus. Departing from thence by sea, they arrived at Acre in the month of April, 1269, and there learned, with extreme concern, that Pope Clement the Fourth was recently dead. A legate whom he had appointed, named M. Tebaldo de' Vesconti di Piacenza, was at this time resident in Acre, and to him they gave an account of what they had in command from the Grand Khan of Tartary. He advised them by all means to wait the election of another Pope, and when that should take place, to proceed with the objects of their embassy. Approving of this counsel, they determined upon employing the interval in a visit to their families in Venice. They accordingly embarked at Acre in a ship bound to Negropont, and from thence went on to Venice, where Nicolo Polo found that his wife, whom he had left with child at his departure, was dead, after having been delivered of a son, who received the name of Macro, and was now of the age of nineteen years. This is the Marco by whom the present work is composed, and who will give therein a relation of all those matters of which he has been an eyewitness.

In the meantime the election of a Pope was retarded by so many obstacles, that they remained two years in Venice, continually expecting its accomplishment; when at length, becoming apprehensive that the Grand Khan might be displeased at their delay, or might suppose it was not their intention to revisit his country, they judged it expedient to return to Acre; and on this occasion they took with them young Macro Polo. Under the sanction of the legate they made a visit to Jerusalem, and there provided themselves with some of the oil belonging to the lamp of the Holy Sepulchre, conforming to the directions of the Grand Khan. As soon as they were furnished with his letters addressed to that prince, bearing testimony to the fidelity with which they had endeavoured to execute his commission, and explaining to him that the Pope of the Christian church had not as yet been chosen, they proceeded to the before-mentioned port of Laiassus. Scarcely however had they taken their departure, when the legate received messengers from Italy, dispatched by the College of Cardinals, announcing his own elevation to the Papal Chair; and he thereupon assumed the name of Gregory the Tenth. Considering that he was now in a situation that enabled him fully to satisfy the wishes of the Tartar sovereign, he hastened to transmit the letters to the king of Armenia, communicating to him the event of his election, and requesting, in case the two ambassadors who were on their way to the court of the Grand Khan should not have already quitted his dominions, that he would give directions for their immediate return. These letters found them still in Armenia, and with great alacrity they obeyed the summons to repair once more to Acre; for which purpose the king furnished them with an armed galley; sending at the same time an ambassador from himself, to offer his congratulations to the Sovereign Pontiff.

Upon their arrival, His Holiness received them in a distinguished manner, and immediately dispatched them with letters papal, accompanied by two friars of the order of Preachers, who happened to be on the spot; men of letters and of science, as well as profound theologians. One of them was named Fra Nicola da Vicenza, and the other,

Fra Guielmo da Tripoli. To them he gave licence and authority to ordain priests, to consecrate bishops, and to grant absolution as fully as he could do in his own person. He also charged them with valuable presents, and among these, several handsome vases of crystal, to be delivered to the Grand Khan in his name, and along with his benediction. Having taken leave, they again steered their course to the port of Laiassus, where they landed, and from thence proceeded into the country of Armenia. Here they received intelligence that the soldan of Babylonia, named Bundokdari, had invaded the Armenian territory with a numerous army, and had overrun and laid waste the country to a great extent. Terrified at these accounts, and apprehensive for their lives, the two friars determined not to proceed further, and delivering over to the Venetians the letters and presents entrusted to them by the Pope, they placed themselves under the protection of the Master of the Knights Templars, and with him returned directly to the coast. Nicolo, Maffeo, and Marco, however, undismayed by perils or difficulties (to which they had long been inured), passed the borders of Armenia, and prosecuted their journey. After crossing deserts of several days' march, and passing many dangerous defiles, they advanced so far, in a direction between northeast and north, that at length they gained information of the Grand Khan, who then had his residence in a large and magnificent city named Clemen-fu. Their whole journey to this place occupied no less than three years and a half; but, during the winter months, their progress had been inconsiderable. The Grand Khan having notice of their approach whilst still remote, and being aware how much they must have suffered from fatigue, sent forward to meet them at the distance of forty days' journey, and gave orders to prepare in every place through which they were to pass, whatever might be requisite to their comfort. By these means, and through the blessing of God, they were conveyed in safety to the royal court.

Upon their arrival they were honourably and graciously received by the Grand Khan, in a full assembly of his principal officers. When they drew nigh to his person, they paid their respects by prostrating themselves on the floor. He immediately commanded them to rise, and to relate to him the circumstances of their travels, with all that had taken place in their negotiation with His Holiness the Pope. To their narrative, which they gave in the regular order of events, and delivered in perspicuous language, he listened with attentive silence. The letters and the presents from Pope Gregory were then laid before him, and, upon hearing the former read, he bestowed much commendation on the fidelity, the zeal, and the diligence of his ambassadors; and receiving with due reverence the oil from the holy sepulchre, he gave directions that it should be preserved with religious care. Upon his observing Marco Polo, and inquiring who he was, Nicolo made answer, "This is your servant, and my son," upon which the Grand Khan replied, "He is welcome, and it pleases me much," and he caused him to be enrolled amongst his attendants of honour. And on account of their return he made a great feast and rejoicing; and as long as the

From *The Book of Ser Marco Polo*.

said brothers and Marco remained in the court of the Grand Khan, they were honoured even above his own courtiers.

Marco was held in high estimation and respect by all belonging to the court. He learnt in a short time and adopted the manners of the Tartars, and acquired proficiency in four different languages, which he became qualified to read and write. Finding him thus accomplished, his master was desirous of putting his talents for business to the proof, and sent him on an important concern of state to a city named Karazan situated at the distance of six months' journey from the imperial residence; on which occasion he conducted himself with so much wisdom and prudence in the management of the affairs entrusted to him, that his services became highly acceptable. On his part, perceiving that the Grand Khan took a pleasure in hearing accounts of whatever was new to him respecting the customs and manners of people, and the peculiar circumstances of distant countries, he endeavoured, wherever he went, to obtain correct information on these subjects, and made notes of all he saw and heard, in order to gratify the curiosity of his master. In short, during seventeen years that he continued in his service, he rendered himself so useful that he was employed on confidential missions to every part of the empire and its dependencies; and sometimes also he travelled on his own private account, but always with the consent, and sanctioned by the authority, of the Grand Khan. Under such circumstances it was that Macro Polo had the opportunity of acquiring a knowledge, either by his own observation, or what he collected from others, of so many things, until his time unknown, respecting the eastern parts of the world, and which he diligently and regularly committed to writing, as in the sequel will appear. And by this means he obtained so much honour, that he provoked the jealousy of the other officers of the court.

Our Venetians having now resided many years at the imperial court, and in that time having realized considerable wealth, in jewels of value and in gold, felt a strong desire to revisit their native country, and, however honoured and caressed by the sovereign, this sentiment was ever predominant in their minds. It became the more decidedly their object, when they reflected on the very advanced age of the Grand Khan, whose death, if it should happen previously to their departure, might deprive them of that public assistance by which alone they could expect to surmount the innumerable difficulties of so long a journey, and reach their homes in safety; which on the contrary, in his lifetime, and through his favour, they might reasonably hope to accomplish. Nicolo Polo accordingly took an opportunity one day, when he observed him to be more than usually cheerful, of throwing himself at his feet, and soliciting on behalf of himself and his family to be indulged with his majesty's gracious permission for their departure. But far from showing himself disposed to comply with the request, he appeared hurt at the application, and asked what motive they could have for wishing to expose themselves to all the inconveniences and hazards of a journey in which they might probably lose their lives. If gain, he said, was their object, he was ready to give them the double of whatever they possessed, and to gratify them with honours to the extent of their desires; but that, from the regard he bore to them, he must positively refuse their petition.

It happened, about this period, that a queen named Bolgana, the wife of Arghun, sovereign of India, died, and as her last request (which she likewise left in a testamentary writing) conjured her husband that no one might succeed to her place on his throne and in his affections, who was not a descendant of her own family, now settled under the dominion of the Grand Khan, in the country of Kathay.

Desirous of complying with this solemn entreaty, Arghun deputed three of his nobles, discreet men, whose names were Ulatai, Apusca, and Goza, attended by a numerous retinue, as his ambassadors to the Grand Khan, with a request that he might receive at his hands a maiden to wife, from among the relatives of his deceased queen. The application was taken in good part, and under the directions of his majesty, choice was made of a damsel aged seventeen, extremely handsome and accomplished, whose name was Kogatin, and of whom the ambassadors, upon her being shown to them, highly approved. When everything was arranged for their departure, and a numerous suite of attendants appointed, to do honour to the future consort of King Arghun, they received from the Grand Khan a gracious dismissal, and set out on their return by the way they came. Having travelled for eight months, their further progress was obstructed and the roads shut up against them, by fresh wars that had broken out amongst the Tartar princes. Much against their inclinations, therefore, they were constrained to adopt the measure of returning to the court of the Grand Khan, to whom they stated the interruption they had met with.

About the time of their reappearance, Macro Polo happened to arrive from a voyage he had made, with a few vessels under his orders, to some parts of the East Indies, and reported to the Grand Khan the intelligence he brought respecting the countries he had visited, with the circumstances of his own navigation, which, he said, was performed in those seas with the utmost safety. This latter observation having reached the ears of the three ambassadors, who were extremely anxious to return to their own country, from whence they had now been absent three years, they presently sought a conference with our Venetians, whom they found equally desirous of revisiting their home; and it was settled between them that the former, accompanied by their young queen, should obtain an audience of the Grand Khan, and represent to him with what convenience and security they might affect their return by sea, to the dominions of their master; while the voyage would be attended with less expense than the journey by land, and be performed in a shorter time; according to the experience of Marco Polo, who had lately sailed in those parts. Should His Majesty incline to give his consent to their adopting that mode of conveyance, they were then to urge him to suffer the three Europeans, as being persons well-skilled in the practice of navigation, to accompany them until they should reach the territory of King Arghun. The Grand Khan upon receiving this application showed by his countenance that it was exceedingly displeasing to him, averse as he was to parting with the Venetians. Feeling nevertheless that he could not with propriety do otherwise than consent, he yielded to their entreaty. Had it not been that he found himself constrained by the importance and urgency of this peculiar case, they would never otherwise have obtained permission to withdraw themselves from his service. He sent for them however, and addressed them with much kindness and condescension, assuring them of his regard, and requiring from them a promise that when they should have resided some time in Europe and with their own family, they would return to him once more. With this object in view he caused them to be furnished with the golden tablet (or royal chop), which contained his order for their having free and safe conduct through every part of his dominions, with the needful supplies for themselves and their attendants. He likewise gave them authority to act in the capacity of his ambassadors to the Pope, the kings of France and Spain, and other Christian princes.

At the same time preparations were made for the equipment of fourteen ships, each having four masts, and capable of being navigated with nine sails, the construction and rigging of which would admit of ample description; but to avoid prolixity, it is for present omitted. Among these vessels there were at least four or five that had crews of two hundred and fifty or two hundred and sixty men. On them were embarked the ambassadors, having the queen under their protection together with Nicolo, Maffeo, and Marco Polo, when they had first taken their leave of the Grand Khan, who presented them with many rubies and other handsome jewels of great value. He also gave directions that the ships should be furnished with stores and provisions for two years.

After a navigation of about three months, they arrived at an island which lay in a southerly direction, named Java, where they saw various objects worthy of attention, of which notice shall be taken in the sequel of the work. Taking their departure from thence, they employed eighteen months in the Indian seas before they were enabled to reach the place of their destination in the territory of King Arghun; and during this part of their voyage also they had an opportunity of observing many things, which shall, in like manner, be related hereafter. But here it may be proper to mention, that between the day of their sailing and that of their arrival, they lost by deaths, of the crews of the vessels and others who were embarked, about six hundred persons; and of the three ambassadors, only one, whose name was Goza, survived the voyage; while of all the ladies and female attendants one only died.

Upon landing they were informed that King Arghun had died some time before, and that the government of the country was then administered, on behalf of his son, who was still a youth, by a person of the name of Ki-akoto. From him they desired to receive instructions as to the manner in which they were to dispose of the princess, whom, by the orders of the late King, they had conducted thither. His answer was that they ought to present the lady to Kasan, the son of Arghun, who was then at a place on the border of Persia, which has its denomination from the Arbor secco, where an army of sixty thousand men was assembled for the purpose of guarding certain passes against the irruption of the enemy. This they proceeded to carry into execution, and having effected it, they returned to the residence of Ki-aka-to, because the road they were afterwards to take lay in that direction. Here, however, they reposed themselves for the space of nine months. When they took their leave he furnished them with four golden tablets, each of them a cubit in length, five inches wide, and weighing three or four marks of gold.

Their inscription began with invoking the blessing of the Almighty upon the Grand Khan, that his name might be held in reverence for many years, and denouncing the punishment of death and confiscation of goods to all who should refuse obedience to the mandate. It then proceeded to direct that the three ambassadors, as his representatives, should be treated throughout his dominions with due honour, that their expenses should be defrayed, and that they should be provided with the necessary escorts. All this was fully complied with, and from many places they were protected by bodies of two hundred horse; nor could this have been dispensed with, as the government of Ki-akato was unpopular, and the people were disposed to commit insults and proceed to outrages, which they would not have dared to attempt under the rule of their proper sovereign. In the course of their journey our travelers received intelligence of the Grand Khan (Kublai) having departed this life; which entirely put an end to all prospect of their revisiting those regions. Pursuing, therefore, their intended route, they at length reached the city of Trebizond, from whence they proceeded to Constantinople, then to Negropont, and finally to Venice, at which place, in the enjoyment of health and abundant riches, they safely arrived in the year 1295. On this occasion they offered up their thanks to God, who had now been pleased to relieve them from such great fatigues, after having preserved them from innumerable perils.

CHAPTER 7

AFRICAN GAME TRAILS

by Theodore Roosevelt

Theodore Roosevelt in Africa. Courtesy of the Library of Congress.

The dangerous game of Africa are the lion, buffalo, elephant, rhinoceros, and leopard. The hunter who follows any of these animals always does so at a certain risk to life or limb; a risk which it is his business to minimize by coolness, caution, good judgment, and straight shooting. The leopard is in point of pluck and ferocity more than the equal of the other four; but his small size always renders it likely that he will merely maul, and not kill, a man. My friend, Carl Akeley, of Chicago, actually killed bare-handed a leopard which sprang on him. He had already wounded the beast twice, crippling it in one front and one hind paw; whereupon it charged, followed him as he tried to dodge the charge, and struck him full just as he turned. It bit him in one arm, biting again and again as it worked up the arm from the wrist to the elbow; but Akeley threw it, holding its throat with the other hand, and flinging its body to one side. It luckily fell on its side with its two wounded legs uppermost, so that it could not tear him. He fell forward with it and crushed in its chest with his knees until he distinctly felt one of its ribs crack; this, said Akeley, was the first moment when he felt he might conquer. Redoubling his efforts, with knees and hands, he actually choked and crushed the life out of it, although his arm was badly bitten. A leopard will charge at least as readily as one of the big beasts, and is rather more apt to get his charge home, but the risk is less to life than to limb.

There are other animals often or occasionally dangerous to human life which are, nevertheless, not dangerous to the hunter. Crocodiles are far greater pests, and far more often man-eaters, than lions or leopards; but their shooting is not accompanied by the smallest element of risk. Poisonous snakes are fruitful sources of accident, but they are actuated only by fear, and the anger born of fear. The hippopotamus sometimes destroys boats and kills those in them; but again there is no risk in hunting him. Finally, the hyena, too cowardly ever to be a source of danger to the hunter, is sometimes a dreadful curse to the weak and helpless. The hyena is a beast of unusual strength, and of enormous power in his jaws and teeth, and thrice over would he be dreaded were fang and sinew driven by a heart of the leopard's cruel courage. But though the creature's foul and evil ferocity has no such backing as that yielded by the angry daring of the spotted cat, it is yet fraught with a terror all its own;

for on occasion the hyena takes to man-eating after its own fashion. Carrion-feeder though it is, in certain places it will enter native huts and carry away children or even sleeping adults; and where famine or disease has worked havoc among a people, hideous spotted beasts become bolder and prey on the survivors. For some years past Uganda has been scourged by the sleeping sickness, which has ravaged it as in the Middle Ages the Black Death ravaged Europe. Hundreds of thousands of natives have died. Every effort has been made by the Government officials to cope with the disease; and among other things sleeping-sickness camps have been established, where those stricken by the dread malady can be isolated and cease to be possible sources of infection to their fellows, Recovery among those stricken is so rare as to be almost unknown, but the disease is often slow, and months may elapse during which the diseased man is still able to live his life much as usual. In the big camps of doomed men and women thus established there were, therefore, many persons carrying on their avocations much as in an ordinary native village. But the hyenas speedily found that in many of the huts the inmates were a helpless prey. In 1908 and throughout the early part of 1909 they grew constantly bolder, haunting these sleeping-sickness camps, and each night entering them, bursting into the huts and carrying off and eating the dying people. To guard against them each little group of huts was enclosed by a thick hedge; but after a while the hyenas learned to break through the hedges, and continued their ravages; so that every night armed sentries had to patrol the camps, and every night they could be heard firing at the marauders.

The men thus preyed on were sick to death, and for the most part helpless. But occasionally men in full vigor are attacked. One of Pease's native hunters was seized by a hyena as he slept beside the campfire, and part of his face torn off. Selous informed me that a friend of his,

Major R. T. Coryndon, then administrator of Northwestern Rhode-
sia, was attacked by a hyena but two or three years ago. At the time
Major Coryndon was lying, wrapped in a blanket, beside his wagon.
A hyena, stealthily approaching through the night, seized him by the
hand, and dragged him out of bed; but as he struggled and called out,
the beast left him and ran off into the darkness. In spite of his torn
hand the major was determined to get his assailant, which he felt sure
would soon return. Accordingly, he went back to his bed, drew his
cocked rifle beside him, pointing toward his feet, and feigned sleep.
When all was still once more, a dim form loomed up through the
uncertain light, toward the foot of the bed; it was the ravenous beast
returning for his prey; and the major shot and killed it where it stood.

A few months ago a hyena entered the outskirts of Nairobi, crept
into a hut, and seized and killed a native man. At Nairobi the wild
creatures are always at the threshold of the town, and often cross it. At
Governor Jackson's table, at Government House, I met Mr. and Mrs.
Sandiford, Mr. Sandiford is managing the railroad. A few months previously, while he was sitting, with his family, in his own house in Nairobi, he happened to ask his daughter to look for something in one of
the bedrooms. She returned in a minute, quietly remarking, "Father,
there's a leopard under the bed." So there was; and it was then remembered that the house-cat had been showing a marked and alert distrust
of the room in question—very probably the leopard had gotten into
the house while trying to catch the cat or one of the dogs. A neighbor
with a rifle was summoned, and shot the leopard.

Hyenas not infrequently kill mules and donkeys, tearing open
their bellies, and eating them while they are still alive. Yet when themselves assailed they usually behave with abject cowardice. The Hills
had a large Airedale terrier, an energetic dog of much courage. Not
long before our visit this dog put up a hyena from a bushy ravine,

in broad daylight, ran after it, overtook it, and flew at it. The hyena made no effective fight, although the dog—not a third its weight—bit it severely, and delayed its flight so that it was killed. During the first few weeks of our trip I not infrequently heard hyenas after nightfall, but saw none. Kermit, however, put one out of a ravine or dry creek-bed—a donga, as it is locally called—and though the brute had a long start he galloped after it and succeeded in running it down. The chase was a long one, for twice the hyena got in such rocky country that he almost distanced his pursuer; but at last, after covering nearly ten miles, Kermit ran into it in the open, shooting it from the saddle as it shambled along at a canter growling with rage and terror. I would not have recognized the cry of the hyenas from what I had read, and it was long before I heard them laugh. Pease said that he had only once

heard them really laugh. On that occasion he was watching for lions outside a Somali zareba. Suddenly a leopard leaped clear over the zareba, close beside him, and in a few seconds came flying back again, over the high thorn fence, with a sheep in its mouth; but no sooner had it landed than the hyenas rushed at it and took away the sheep; and then their cackling and shrieking sounded exactly like the most unpleasant kind of laughter. The normal death of very old lions, as they

Roosevelt with one of his quarry—a crocodile.

grow starved and feeble—unless they are previously killed in an encounter with dangerous game like buffalo—is to be killed and eaten by hyenas; but of course a lion in full vigor pays no heed to hyenas, unless it is to kill one if it gets in the way.

During the last few decades, in Africa, hundreds of white hunters, and thousands of native hunters, have been killed or wounded by lions, buffaloes, elephants, and rhinos. All are dangerous game; each species has to its gruesome credit a long list of mighty hunters slain or disabled. Among those competent to express judgment there is the widest difference of opinion as to the comparative danger in hunting the several kinds of animals. Probably no other hunter who has ever lived has combined Selous's experience with his skill as a hunter and his power of accurate observation and narration. He has killed between three and four hundred lions, elephants, buffaloes, and rhinos, and he ranks the lion as much the most dangerous, and the rhino as much the least, while he puts the buffalo and elephant in between, and practically on a par. Governor Jackson has killed between eighty and ninety of the four animals; and he puts the buffalo unquestionably first in point of formidable capacity as a foe, the elephant equally unquestionably second, the lion third, and the rhino last. Stigand puts them in the following order: lion, elephant, rhino, leopard, and buffalo. Drummond, who wrote a capital book on South African game, who was for years a professional hunter like Selous, and who had fine opportunities for observation, but who was a much less accurate observer than Selous, put the rhino as unquestionably the most dangerous, with the lion as second, and the buffalo and elephant nearly on a level. Samuel Baker, a mighty hunter and good observer, but with less experience of African game than any one of the above, put the elephant first, the rhino second, the buffalo seemingly third, and the lion last. The experts of greatest experience thus absolutely disagree among themselves; and there is the same wide divergence of view among good hunters and trained observers whose opportunities have been less. Mr. Abel Chapman, for instance, regards both the elephant and the rhino as more dangerous than the lion; and many of the hunters I met in East Africa seemed inclined to rank the buffalo as more dangerous than any other animal. A man who has shot but a dozen or a score of these various animals, all put together, is not entitled to express any but the most tentative opinion as to their relative prowess and ferocity; yet on the whole it seems to me that the weight of opinion among those best fitted to judge is that the lion is the most formidable opponent of the hunter, under ordinary conditions. This is my own view. But we must ever keep in mind the fact that the surrounding conditions, the geographical locality, and the wide individual variation of temper within the ranks of each species, must all be taken into account. Under certain circumstances a lion may be easily killed, whereas a rhino would be a dangerous foe. Under other conditions the rhino could be attacked with impunity, and the lion only with the utmost hazard; and one bull buffalo might flee and one bull elephant charge, and yet the next couple met with might show an exact reversal of behavior.

At any rate, during the last three or four years, in German and British East Africa and Uganda, over fifty white men have been killed or mauled by lions, buffaloes, elephants, and rhinos; and the lions have much the largest list of victims to their credit. In Nairobi churchyard I was shown the graves of seven men who had been killed by lions, and of one who had been killed by a rhino. The first man to meet us on the African shore was Mr. Campbell, Governor Jackson's A.D.C., and only a year previously he had been badly mauled by a lion. We met one gentleman who had been crippled for life by a lioness. He had marked her into some patches of brush, and coming up, tried to put her out of one thick clump. Failing, he thought she might have gone into another thicket, and walked toward it; instantly that his back was turned, the lioness, who had really been in the first clump of brush, raced out after him, threw him down, and bit him again and again before she was driven off. One night we camped at the very spot where, a score of years before, a strange tragedy had happened. It was in the early days of the opening of the country, and an expedition was going toward Uganda; one of the officials in charge was sleeping in a tent with the flap open. There was an askari on duty; yet a lion crept up, entered the tent, and seized and dragged forth the man. He struggled and made outcry; there was a rush of people, and the lion dropped his prey and bounded off. The man's wounds were dressed, and he was put back to bed in his own tent; but an hour or two after the camp again grew still, the lion returned, bent on the victim of whom he had been robbed; he re-entered the tent, seized the unfortunate wounded man with his great fangs, and this time made off with him into the surrounding darkness, killed and ate him. Not far from the scene of this tragedy, another had occurred. An English officer named Stewart, while endeavoring to kill his first lion, was himself set on and slain. At yet another place we were shown were two settlers, Messrs. Lucas and Goldfinch, had been one killed and one crippled by a lion they had been hunting. They had been following the chase on horseback, and being men of bold nature, and having killed several lions, had become too daring. They hunted the lion into a small piece of brush and rode too near it. It came out at a run and was on them before their horses could get under way. Goldfinch was knocked over and badly bitten and clawed; Lucas went to his assistance, and was in his turn knocked over, and the lion then lay on him and bit him to death. Goldfinch, in spite of his own severe wounds, crawled over and shot the great beast as it lay on his friend.

Most of the settlers with whom I was hunting had met with various adventures in connection with lions. Sir Alfred had shot many in different parts of Africa; some had charged fiercely, but he always stopped them. Captain Slatter had killed a big male with a mane a few months previously. He was hunting it in company with Mr. Humphery, the District Commissioner of whom I have already spoken, and it gave them some exciting moments, for when hit it charged savagely. Humphery had a shotgun loaded with buckshot, Slatter his rifle. When wounded, the lion charged straight home, hit Slatter, knocking him flat and rolling him over and over in the sand, and then went after the native gun-bearer, who was running away—the worst possible course to follow with a charging lion. The mechanism of Setter's rifle was choked by the sand, and as he rose to his feet he saw the lion overtake the fleeing man, rise on his hind legs like a rearing horse—not springing—and strike down the fugitive. Humphery fired into him with buckshot, which merely went through the skin; and some minutes elapsed before Slatter was able to get his rifle in shape to kill the lion, which, fortunately, had begun to feel the effect of his wounds, and was too sick to resume hostilities of its own accord. The gun-bearer was badly but not fatally injured. Before this, Slatter, while on a lion hunt, had been set afoot by one of the animals he was after, which had killed his horse. It was at night and the horse was tethered within six yards of his sleeping master. The latter was aroused by the horse galloping off, and he heard it staggering on for some sixty yards before it fell. He and his friend followed it with lanterns and drove off the lion, but the horse was dead. The tracks and the marks on the horse showed what had happened. The lion had sprung clean on the horse's back, his foreclaws dug into the horse's shoulders, his hind claws cutting into its haunches, while the great fangs bit at the neck. The horse struggled off at a heavy run, carrying its fearsome burden. After going some sixty yards the lion's teeth went through the spinal cord, and the ride was over. Neither animal had made a sound and the lion's feet did not touch the earth until the horse fell.

While a magistrate in the Transvaal, Pease had under him as game officer a white hunter, a fine fellow, who underwent an extraordinary experience. He had been off some distance with his Kaffir boys, to hunt a lion. On his way home the hunter was hunted. It was after nightfall. He had reached a region where lions had not been seen for a long time, and where an attack by them was unknown. He was riding along a trail in the darkness, his big boarhound trotting ahead, his native "boys" some distance behind. He heard a rustle in the bushes alongside the path, but paid no heed, thinking it was a reedbuck. Immediately afterward two lions came out in the path behind and raced after him. One sprang on him, tore him out of the saddle, and trotted off holding him in its mouth, while the other continued after the frightened horse. The lion had him by the right shoulder, and yet with his left hand he wrenched his knife out of his belt and twice stabbed it. The second stab went to the heart and the beast let go of him, stood a moment, and fell dead. Meanwhile, the dog had followed the other lion, which now, having abandoned the chase of the horse, and with the dog still at his heels, came trotting back to look for the man. Crippled though he was, the hunter managed to climb a small tree; and though the lion might have gotten him out of it, the dog interfered. Whenever the lion came toward the tree the dog worried him, and kept him off until, at the shouts and torches of the approaching Kaffir boys, he sullenly retired, and the hunter was rescued.

Percival had a narrow escape from a lion, which nearly got him, though probably under a misunderstanding. He was riding through a wet spot of ground, where the grass was four feet high, when his horse suddenly burst into a run and the next moment a lion had galloped almost alongside of him. Probably the lion thought it was a zebra, for when Percival, leaning over, yelled in his face, the lion stopped short. But he at once came on again, and nearly caught the horse. However, they were now out of the tall grass, and the lion gradually drew up when they reached the open country.

The two Hills, Clifford and Harold, were running an ostrich farm. The lions sometimes killed their ostriches and stock; and the Hills in return had killed several lions. The Hills were fine fellows; Africanders, as their forefathers for three generations had been, and frontiersmen of the best kind. From the first moment they and I became fast friends, for we instinctively understood one another, and found that we felt alike on all the big questions, and looked at life, and especially the life of effort led by the pioneer settler, from the same standpoint. They reminded me, at every moment, of those Western ranchmen and homemakers with whom I have always felt a special sense of companionship and with whose ideals and aspirations I have always felt a special sympathy. A couple of months before my visit, Harold Hill had met with a rather unpleasant adventure. He was walking home across the lonely plains, in the broad daylight, never dreaming that lions might be abroad, and was unarmed. When still some miles from his house while plodding along, he glanced up and saw three lions in the trail only fifty yards off, staring fixedly at him. It happened to be a place where the grass was rather tall, and lions are always bold where there is the slightest cover; whereas, unless angered, they are cautious on bare ground. He halted, and then walked slowly to one side; and then slowly forward toward his house. The lions followed him with their eyes, and when he had passed they rose and slouched after him. They were not pleasant followers, but to hurry would have been fatal; and he walked slowly on along the road, while for a mile he kept catching glimpses of the tawny bodies of the beasts as they trod stealthily forward through the sunburned grass, alongside or a little behind him. Then the grass grew short, and the lions halted and continued to gaze after him until he disappeared over a rise.

Everywhere throughout the country we were crossing were signs that the lion was lord and that his reign was cruel. There were many lions, for the game on which they feed was extraordinarily abundant. They occasionally took the ostriches or stock of the settlers, ravaged the herds and flocks of the natives, but not often; for their favorite food was yielded by the swarming herds of kongoni and zebras, on which they could prey at will. Later we found that in this region they rarely molested the buffalo, even where they lived in the same reed-beds; and this though elsewhere they habitually prey on the buffalo. But where zebras and harte-beests could be obtained without effort, it was evidently not worth their while to challenge such formidable quarry. Every "kill" I saw was a kongoni or a zebra; probably I came across fifty of each. One zebra kill, which was not more than eighteen hours old (after the lapse of that time the vultures and marabouts, not to speak of the hyenas and jackals, leave only the bare bones), showed just what had occurred. The bones were all in place, and the skin still on the lower legs and head. The animal was lying on its belly, the legs spread out, the neck vertebra crushed; evidently the lion had sprung clean on it, bearing it down by his weight while he bit through the back of the neck, and the zebra's legs had spread out as the body yielded under the lion. One fresh kongoni kill showed no marks on the haunches, but a broken neck and claw marks on the face and withers; in this case the lion's hind legs had remained on the ground, while with his fore paws he grasped the kongoni's head and shoulders, holding it until the teeth splintered the neck bone.

One or two of our efforts to get lions failed, of course; the ravines we beat did not contain them, or we failed to make them leave some particularly difficult hill or swamp—for lions lie close. But Sir Alfred knew just the right place to go to, and was bound to get us lions—and he did.

One day we started from the ranch house in good season for an all-day lion hunt. Besides Kermit and myself, there was a fellow guest, Medlicott, and not only our host, but our hostess and her daughter; and we were joined by Percival at lunch, which we took under a great fig-tree, at the foot of a high, rocky hill. Percival had with him a little mongrel bull-dog, and a Masai "boy," a fine, bold-looking savage, with a handsome head-dress and the usual formidable spear; master, man, and dog evidently all looked upon any form of encounter with lions simply in the light of a spree.

After lunch we began to beat down a long donga, or dry watercourse—a creek, as we should call it in the Western plains country. The watercourse, with low, steep banks, wound in curves, and here and there were patches of brush, which might contain anything in the shape of lion, cheetah, hyena, or wild dog. Soon we came upon lion spoor in the sandy bed; first the footprints of a big male, then those of a lioness. We walked cautiously along each side of the donga, the horses following close behind so that if the lion were missed we could gallop after him and round him up on the plain. The dogs—for besides the little bull, we had a large brindled mongrel named Ben, whose courage belied his looks—began to show signs of scenting the lion; and we beat out each patch of brush, the natives shouting and throwing in stones, while we stood with the rifles where we could best command any probable exit. After a couple of false alarms the dogs drew toward one patch, their hair bristling, and showing such eager excitement that it was evident something big was inside; and in a moment one of the boys called, "samba" (lion), and pointed with his finger. It was just across the little ravine, there about four yards wide and as many feet deep; and I shifted my position, peering eagerly into the bushes for some moments before I caught a glimpse of tawny hide; as it moved, there was a call to me to "shoot," for at that distance, if the lion charged, there would be scant time to stop it; and I fired into what I saw. There was a commotion in the bushes, and Kermit fired; and immediately afterward there broke out on the other side, not the hoped-for big lion, but two cubs the size of mastiffs. Each was badly wounded and we finished them off; even if unwounded they were too big to take alive.

This was a great disappointment, and as it was well on in the afternoon, and we had beaten the country most apt to harbor our game, it seemed unlikely that we would have another chance. Percival was on foot and a long way from his house, so he started for it; and the rest of us also began to jog homeward. But Sir Alfred, although he said nothing, intended to have another try. After going a mile or two he started off to the left at a brisk canter; and we, the other riders, followed, leaving behind our gun-bearers, saises, and porters. A couple of miles away was another donga, another shallow watercourse with occasional big brush patches along the winding bed; and toward this we cantered. Almost as soon as we reached it our leader found the spoor of two big lions; and with every sense acock, we dismounted and approached the first patch of tall bushes. We shouted and threw in stones, but nothing came out; and another small patch showed the same result. Then we mounted our horses again, and rode toward another patch a quarter of a mile off. I was mounted on Tranquillity, the stout and quiet sorrel.

This patch of tall, thick brush stood on the hither bank—that is, on our side of the watercourse. We rode up to it and shouted loudly. The response was immediate, in the shape of loud gruntings and crashings through the thick brush. We were off our horses in an instant, I throwing the reins over the head of mine; and without delay the good old fellow began placidly grazing, quite unmoved by the ominous sounds immediately in front.

I sprang to one side; and for a second or two we waited, uncertain whether we should see the lions charging out ten yards distant or running away. Fortunately, they adopted the latter course. Right in front of me, thirty yards off, there appeared, from behind the bushes which had fist screened him from my eyes, the tawny, galloping form of a big maneless lion. Crack! The Winchester spoke; and as the soft-nosed bullet ploughed forward through his flank the lion swerved so that I missed him with the second shot; but my third bullet went through the spine and forward into his chest, down he came, sixty yards off, his hind quarters dragging, his head up, his ears back, his jaws open and lips drawn up in a prodigious snarl, as he endeavored to turn to face us. His back was broken; but of this we could not at the moment be sure, and if it had merely been grazed, he might have recovered, and then, even though dying, his charge might have done mischief. So Kermit, Sir Alfred, and I fired, almost together, into his chest. His head sank, and he died.

This lion had come out on the left of the bushes; the other, to the right of them, had not been hit, and we saw him galloping off across the plain, six or eight hundred yards away. A couple more shots missed, and we mounted our horses to try to ride him down. The plain sloped gently upward for three-quarters of a mile to a low crest or divide, and long before we got near him he disappeared over this. Sir Alfred and Kermit were tearing along in front and to the right, with Miss Pease close behind; while Tranquillity carried me, as fast as he could, on the left, with Medlicott near me. On topping the divide Sir Alfred and Kermit missed the lion, which had swung to the left, and they raced ahead too far to the right. Medlicott and I, however, saw the lion, loping along close behind some kongoni; and this enabled me to get up to him as quickly as the lighter men on the faster horses. The going was now slightly downhill, and the sorrel took me along very well, while Medlicott, whose horse was slow, bore to the right and joined the other two men. We gained rapidly, and, finding out this, the lion suddenly halted and came to bay in a slight hollow, where the grass was rather long. The plain seemed flat, and we could see the lion well from horseback; but, especially when he lay down, it was most difficult to make him out on foot, and impossible to do so when kneeling.

We were about a hundred and fifty yards from the lion, Sir Alfred, Kermit, Medlicott, and Miss Pease off to one side, and slightly above him and them. Kermit and I tried shooting from the horses; but at such a distance this was not effective. Then Kermit got off, but his

"Hunting the Lion" by Frederic William Unger, 1909. Courtesy of the Smithsonian Institute.

horse would not let him shoot; and when I got off I could not make out the animal through the grass with sufficient distinctness to enable me to take aim. Old Ben the dog had arrived, and barking loudly, was strolling about near the lion, which paid him not the slightest attention. At this moment my black sais, Simba, came running up to me and took hold of the bridle; he had seen the chase from the line of march and had cut across to join me. There was no other sais or gun-bearer anywhere near, and his action was plucky, for he was the only man afoot, with the lion at bay. Lady Pease had also ridden up and was an interested spectator only some fifty yards behind me.

Now, an elderly man with a varied past which includes rheumatism does not vault lightly into the saddle; as his sons, for instance, can; and I had already made up my mind that in the event of the lion's charging it would be wise for me to trust to straight powder rather than to try to scramble into the saddle and get under way in time. The arrival of my two companions settled matters. I was not sure of the speed of Lady Pease's horse; and Simba was on foot and it was of course out of the question for me to leave him. So I said, "Good, Simba, now we'll see this thing through," and gentle-mannered Simba smiled a shy appreciation of my tone, though he could not understand the words. I was still unable to see the lion when I knelt, but he was now standing up, looking first at one group of horses and then at the other, his tail lashing to and fro, his head held low, and his lips dropped over his mouth in peculiar fashion, while his harsh and savage growling rolled thunderously over the plain. Seeing Simba and me on foot, he turned toward us, his tail lashing quicker and quicker. Resting my elbow on Simba's bent shoulder, I took steady aim and pressed the trigger, the bullet went in between the neck and shoulder, and the lion fell over on his side, one foreleg in the air. He recovered in a moment and stood up, evidently very sick, and once more faced me, growling hoarsely.

I think he was on the eve of charging. I fired again at once, and this bullet broke his back just behind the shoulders; and with the next I killed him outright, after we had gathered round him.

These were two good-sized maneless lions; and very proud of them I was. I think Sir Alfred was at least as proud, especially because we had performed the feat alone, without any professional hunters being present, "We were all amateurs, only gentlemen riders up," said Sir Alfred. It was late before we got the lions skinned. Then we set off toward the ranch, two porters carrying each lion skin, strapped to a pole; and two others carrying the cub skins. Night fell long before we were near the ranch; but the brilliant tropic moon lighted the trail. The stalwart savages who carried the bloody lion skins swung along at a faster walk as the sun went down and the moon rose higher; and they began to chant in unison, one uttering a single word or sentence, and the others joining in a deep-toned, musical chorus. The men on a safari, and indeed African natives generally, are always excited over the death of a lion, and the hunting tribes then chant their rough hunting songs, or victory songs, until the monotonous, rhythmical repetitions make them grow almost frenzied. The ride home through the moonlight, the vast barren landscape shining like silver on either hand, was one to be remembered; and above all, the sight of our trophies and of their wild bearers.

Three days later we had another successful lion hunt. Our camp was pitched at a waterhole in a little stream called Potha, by a hill of the same name. Pease, Medlicott, and both the Hills were with us, and Heller came too; for he liked, when possible, to be with the hunters so that he could at once care for any beast that was shot. As the safari was stationary, we took fifty or sixty porters as beaters. It was thirteen hours before we got into camp that evening. The Hills had with them as beaters and water-carriers half a dozen of the Wakamba who were working on their farm. It was interesting to watch these naked savages, with their filed teeth, their heads shaved in curious patterns, and carrying for arms little bows and arrows.

Before lunch we beat a long, low hill. Harold Hill was with me; Medlicott and Kermit were together. We placed ourselves, one couple on each side of a narrow neck, two-thirds of the way along the crest of the hill; and soon after we were in position we heard the distant shouts of the beaters as they came toward us, covering the crest and the tops of the slopes on both sides. It was rather disconcerting to find how much better Hill's eyes were than mine. He saw everything first, and it usually took some time before he could make me see it. In this first drive nothing came my way except some mountain reedbuck does, at which I did not shoot. But a fine male cheetah came to Kermit, and he bowled it over in good style as it ran.

Then the beaters halted, and waited before resuming their march until the guns had gone clear round and established themselves at the base of the farther end of the hill. This time Kermit, who was a couple of hundred yards from me, killed a reedbuck and a steinbuck. Suddenly Hill said "Lion," and endeavored to point it out to me, as it crept cautiously among the rocks on the steep hillside, a hundred and fifty yards away. At first I could not see it; finally I thought I did and fired, but, as it proved, at a place just above him. However, it made him start up, and I immediately put the next bullet behind his shoulders; it was a fatal shot; but, growling, he struggled down the hill, and I fired again and killed him. It was not much of a trophy, however, turning out to be a half-grown male.

We lunched under a tree, and then arranged for another beat. There was a long, wide valley, or rather a slight depression in the ground—for it was only three or four feet below the general level—in which the grass grew tall, as the soil was quite wet. It was the scene of Percival's adventure with the lion that chased him. Hill and I stationed ourselves on one side of this valley or depression, toward the upper end; Pease took Kermit to the opposite side; and we waited, our horses some distance behind us. The beaters were put in at the lower end, formed a line across the valley, and beat slowly toward us, making a great noise.

They were still some distance away when Hill saw three lions, which had slunk stealthily off ahead of them through the grass. I have called the grass tall, but this was only by comparison with the short grass of the dry plains. In the depression or valley it was some three feet high. In such grass a lion, which is marvellously adept at hiding, can easily conceal itself, not merely when lying down, but when advancing at a crouching gait. If it stands erect, however, it can be seen.

There were two lions near us, one directly in our front, a hundred and ten yards off. Some seconds passed before Hill could make me realize that the dim yellow smear in the yellow-brown grass was a lion; and then I found such difficulty in getting a bead on him that I overshot. However, the bullet must have passed very close—indeed, I think it just grazed him—for he jumped up and faced us, growling savagely. Then, his head lowered, he threw his tail straight into the air and began to charge. The first few steps he took at a trot, and before he could start into a gallop I put the soft-nosed Winchester bullet in between the neck and shoulder. Down he went with a roar; the wound was fatal, but I was taking no chances, and I put two more bullets in him. Then we walked toward where Hill had already seen another lion—the lioness, as it proved. Again he had some difficulty in making me see her; but he succeeded and I walked toward her through the long grass, repressing the zeal of my two gun-bearers, who were stanch, but who showed a tendency to walk a little ahead of me on each side, instead of a little behind. I walked toward her because I could not kneel to shoot in grass so tall; and when shooting off-hand I like to be fairly close, so as to be sure that my bullets go in the right place. At sixty yards I could make her out clearly, snarling at me as she faced me; and I shot her full in the chest. She at once performed a series of extraordinary antics, tumbling about on her head, just as if she were throwing somersaults, first to one side and then to the other. I fired again, but managed to shoot between the somersaults, so to speak, and missed her. The shot seemed to bring her to herself, and away she tore; but instead of charging us she charged the line of beaters. She was dying fast, however, and in her weakness failed to catch anyone; and she sank down into the long grass. Hill and I advanced to look her up, our rifles at full cock, and the gun-bearers close behind. It is ticklish work to follow a wounded lion in tall grass, and we walked carefully, every sense on the alert. We passed Heller, who had been with the beaters. He spoke to us with an amused smile. His only weapon was a pair of field-glasses, but he always took things as they came, with entire coolness, and to be close to a wounded lioness when she charged merely interested him. A beater came running up and pointed toward where he had seen her, and we walked toward the place. At thirty yards' distance Hill pointed, and, eagerly peering, I made out the form of the lioness showing indistinctly through the grass. She was half crouching, half sitting, her head bent down; but she still had strength to do mischief. She saw us, but before she could turn I sent a bullet through her shoulders; down she went, and was dead when we walked up. A cub had been seen, and another full-grown lion, but they had slunk off and we got neither.

This was a full-grown, but young, lioness of average size; her cubs must have been several months old. We took her entire to camp to weigh; she weighed two hundred and eighty-three pounds. The first lion, which we had difficulty in finding, as there were no identifying marks in the plain of tall grass, was a good-sized male, weighing about four hundred pounds, but not yet full-grown; although he was probably the father of the cubs.

We were a long way from camp, and, after beating in vain for the other lion, we started back; it was after nightfall before we saw the campfires. It was two hours later before the porters appeared, bearing on poles the skin of the dead lion, and the lioness entire. The moon was nearly full, and it was interesting to see them come swinging down the trail in the bright silver light, chanting in deep tones, over and over again, a line or phrase that sounded like:

"Zou-zou-boule ma ja guntai; zou-zou-boule ma ja guntai."

Occasionally they would interrupt it by the repetition in unison, at short intervals, of a guttural ejaculation, sounding like "huzlem." They marched into camp, then up and down the lines, before the rows of small fires; then, accompanied by all the rest of the porters, they paraded up to the big fire where I was standing. Here they stopped and ended the ceremony by a minute or two's vigorous dancing amid singing and wild shouting. The firelight gleamed and flickered across the grim dead beasts and the shining eyes and black features of the excited savages, while all around the moon flooded the landscape with her white light.

CHAPTER 8

HUNTING MUSK-OXEN NEAR THE POLE

by Lieutenant Robert E. Perry, USN

Musk oxen.

On the fifteenth of May, 1895, the storm ceased which had held Lee, Hensen, and myself prisoners for two days upon the Independence Bay moraine, the northern shore of the "Great Ice," more than four thousand feet above the level of the sea. Then, in a very short time, I completed all the preparations for a trip down over the land in search of the musk oxen which would be our salvation. Matt and all the dogs were to accompany me.

I took the little "Chopsie" sledge, our rifles, four days' supply of tea, biscuits, and oil—we had had no meat for several days—and the remainder of the dog food, which was a lump of frozen walrus meat somewhat larger than a man's head. Lee was to remain at the tent during our absence.

The almost total lack of snow on this northern land was a surprise as well as an annoyance to me, since it threatened to damage my sledges seriously. But by keeping well ahead of the dogs, I was able to pick out a fairly good though circuitous path along the numerous snow drifts which lay on the leeward side of the hills and mountains.

After some twelve hours of steady marching, we were close to Musk Ox Valley, where, three years before, Astrup and myself had first seen and killed some of these animals.

Leaving Matt with the sledge and dogs, I took my rifle and entered the valley, hoping to find them there again. So far we had not seen the slightest indication of musk oxen though we had followed the same route, where, on my previous visit, their traces had been visible on almost every square rod of ground.

In the valley I found no trace of their presence, and I returned to the sledge in a gloomy mood. Could it be that the musk oxen of this region were migratory? Did they retreat southward along the east coast in the autumn, and return in early summer, and were we too early for them? Or had the sight and smell of us and our dogs, and the sound of our guns and the sight of carcasses of the oxen we had slain three years before, terrified the others so that they had deserted this region completely?

These questions disturbed me deeply. We had now been marching for a long time; we were tired with the unaccustomed work of climbing up and down hills, and were weak and hungry from our long and scant diet of tea and biscuits.

Our hunger was partially appeased by the dog food. True, this was a frozen mixture of walrus meat, blubber, hair, sand, and various other foreign substances, but we had to eat something, and the fact that the meat was "high" and the blubber more or less rancid did not deter us. Yet we dared not satisfy ourselves, even with dog food, as we were dependent on the dogs, and they were more in need of food than we.

A few miles beyond Musk Ox Valley I saw a fresh hare-track leading in the same direction in which we were going; and within five minutes I saw the hare itself squatting among the rocks a few paces distant. I called to Matt, who was some little distance back, to stop the dogs and come up with his rifle.

He was so affected by the prospect of a good dinner that his first and second bullets missed the mark, although usually he was a good marksman. But at the third shot the beautiful, spotless little animal collapsed into a shapeless mass, and on the instant gaunt hunger took from us the power of further endurance. We must stop at once and cook the hare.

Near us was a little pond surrounded by high banks. This offered the advantages of ice, from which to melt water for cooking purposes. So here we camped, lit our lamp, and cooked and ate the entire hare. It was the first full meal we had had for nearly six weeks—the first meal furnishing sufficient substance and nourishment for the doing of a heavy day's work.

Ease from hunger and pleasure in the process of digestion made us drowsy. Lying down as we were, upon the snow-covered shore of the little pond, without tent or sleeping bag or anything except the clothes we wore, we slept the sleep of tired children in their cosy beds, though the snowflakes were falling thickly upon us.

The next morning we pushed on for a valley near Navy Cliff, where Astrup and myself had seen numerous musk ox tracks. At the entrance of this valley I came upon a track, so indistinct that it might have been made the previous autumn. Following it a short distance, I saw the accompanying tracks of a calf, showing at once that the tracks were of this season; and a little farther on there were traces but a few days old.

Fastening our dogs securely to a rock and muzzling them so that they could neither gnaw themselves loose or make a noise to disturb the musk oxen, we passed rapidly down the valley, Winchesters in hand, with our eyes fixed eagerly upon the tracks. Soon we reached the feeding ground of the herd on the preceding day, and knew by their tracks and the places where they had dug away the snow in search of grass and moss that there was quite a herd of them.

We circled the feeding ground as rapidly as we could, and at length found the tracks of the herd leading out of the labyrinth and up the slopes of the surrounding mountains. Following these, our eyes were soon gladdened by sight of a group of black spots on a little terrace just below the crest of the mountain.

Looking through the field glass, we saw that some of the animals were lying down. Evidently the herd was beginning its midday snooze. We moved cautiously up to the edge of the terrace to leeward of the animals and sought shelter behind a big boulder. The musk oxen were about two hundred yards distant and numbered twenty-two.

I wonder if one of my readers knows what hunger is. Hensen and I were worn to the bone with scant rations and hard work, which had left little on our bones except lean, tense muscles and wires of sinew. The supper from the hare—that meal of fresh, hot, luscious meat— the first adequate meal in nearly six hundred miles of snowshoeing, had wakened every merciless hunger-pang that during the previous weeks had been gradually dulled into insensibility.

Gazing on the big black animals before us, we saw not game, but meat; and every nerve and fibre in my gaunt body was vibrating with a furious and savage hunger for that meat—meat that should be soft and warm; meat into which the teeth could sink and tear and rend; meat that would not blister the lips and tongue with its frost, nor ring like rock against the teeth.

Panting and quivering with excitement we lay for a few moments. We could not risk a shot at that distance.

"Do you think they will come for us?" said Matt.

"God knows I hope so, boy, for then we are sure of some of them. Are you ready?"

"Yes, sir."

"Come on, then."

Rising one of us on one side of the boulder, the other on the other side, we dashed across the rocks and snow straight toward them.

There was a snort and stamp from the big bull guarding the herd, and the next instant every animal was on his feet, facing us, thank God! The next moment they were in close line, with lowered heads and horns. I could have yelled for joy if I had had the breath to spare.

Every one of us has read thrilling stories of deer chased by hungry wolves, and it is in the nature of man to sympathise with the creatures that are trying to escape. But did any of us ever stop to think how those other poor creatures, the wolves, were feeling? I know now just what their feelings are, and I cannot help sympathising with them. I was no better than a wolf myself at that moment.

We were within less than fifty yards of the herd when the big bull with a quick motion lowered his horns still more. Instinct, Providence—call it what you will—told me it was the signal for the herd to charge. Without slackening my pace, I pulled my Winchester to my shoulder and sent a bullet at the back of his neck over the white, impervious shield of the great horns.

Heart and soul and brain and eyes went with that singing bullet. I felt that I was strong enough, and hungry enough, and wild enough so that, had the bull been alone, I could have sprung upon him bare-handed and somehow made meat of him. But against the entire herd we should have been powerless; once the black avalanche had gained momentum, we should have been crushed by it like the crunching snow crystals under our feet.

As the bull fell upon his knees the herd wavered. A cow half turned and, as Matt's rifle cracked, fell with a bullet back of her fore-shoulder. Without raising my rifle above my hips another one dropped. Then another, for Matt; then the herd broke, and we hurried in pursuit.

A wounded cow wheeled and, with lowered head, was about to charge me; again Matt's rifle cracked, and she fell. As I rushed past her he shouted, "That was my last cartridge!"

A short distance beyond, the remainder of the herd faced about again and I put a bullet into the breast of another bull, but it did not stop him, and the herd broke again and disappeared over a sharp ridge. I had neither wind nor strength to follow.

Suddenly the back of one of the animals appeared above the ridge. I whirled and fired. I did not see the sights—I think I scarcely saw my rifle, but felt my aim as I would with harpoon or stone. I heard the thud of the bullet. I knew the beast was hit behind the fore-shoulder. As the animal disappeared I sank down on the snow, quite unable to go farther.

But after a little rest I was able to move. Matt came up. We instantly set about bleeding the last beast I had killed. How delicious that tender, raw, warm meat was—a mouthful here and a mouthful there, cut from the animal as we skinned it! It seems dreadful and loathsome to have made such a meal. But I wish those who stay at home at ease to realise what hunger drives men to. This was the barbarism out of which our race has risen. Matt and I were savages for the time. I ate till I dared eat no more, although still unsatisfied.

Then Matt went back to bring up the dogs and sledge, while I began removing the skins from our game. With Matt's return came the supremest luxury of all! That was to toss great lumps of meat to the gaunt shadows which we called dogs, till they, too, could eat no more, and lay gorged and quiet upon the rocks.

The removal of the great shaggy, black pelts of the musk oxen was neither an easy nor a rapid job. By the time it was completed it was midnight; the sun was low over the mountains in the north, and a biting wind whistled about our airy location.

We were glad to drag the skins to a central place, construct a wind guard with the assistance of the sledge, a few stones, and a couple of the skins, and make a bed of the other on the lee side of it.

We built up a little stone shelter for our cooking lamp, and then, stretched upon our luxurious, thick, soft, warm couch, we were, for the first time, able to spare the time to make ourselves some tea, and cook some of the delicious musk ox meat.

Then, with the savage, sombre northern land lying like a map below us—the barren rocks, mottled here and there with eternal snowdrifts; with the summits of the distant mountains disappearing in a mist of driving snow; with the biting breath of the "Great Ice" following us even here and drifting the fine snow over and about our shelter, we slept again as tired children—nay as tired savages—sleep.

PART THREE

ADVENTURES BY FICTION

CHAPTER 9

"THE TOWN-HO'S STORY" FROM *MOBY DICK*

by Herman Melville

The Cape of Good Hope, and all the watery region round about there, is much like some noted four corners of a great highway, where you meet more travellers than in any other part.

It was not very long after speaking the Goney that another homeward-bound whaleman, the Town-Ho, was encountered. She was manned almost wholly by Polynesians. In the short game that ensued she gave us strong news of Moby Dick. To some the general interest in the White Whale was now wildly heightened by a circumstance of the Town-Ho's story, which seemed obscurely to involve with the whale a certain wondrous, inverted visitation of one of those so called judgments of God which at times are said to overtake some men. This latter circumstance, with its own particular accompaniments, forming what may be called the secret part of the tragedy about to be narrated, never reached the ears of Captain Ahab or his mates. For that secret part of the story was unknown to the captain of the Town-Ho himself. It was the private property of three confederate white seamen of that ship, one of whom, it seems, communicated it to Tashtego with Romish injunctions of secresy, but the following night Tashtego rambled in his sleep, and revealed so much of it in that way, that when he was wakened he could not well withhold the rest. Nevertheless, so potent an influence did this thing have on those seamen in the Pequod who came to the full knowledge of it, and by such a strange delicacy, to call it so, were they governed in this matter, that they kept the secret among themselves so that it never transpired abaft the Pequod's mainmast. Interweaving in its proper place this darker thread with the story as publicly narrated on the ship, the whole of this strange affair I now proceed to put on lasting record.

For my humor's sake, I shall preserve the style in which I once narrated it at Lima, to a lounging circle of my Spanish friends, one Saint's eve, smoking upon the thick-gilt tiled piazza of the Golden Inn. Of those fine cavaliers, the young Dons, Pedro and Sebastian, were on the closer terms with me; and hence the interluding questions they occasionally put, and which are duly answered at the time.

"Some two years prior to my first learning the events which I am about rehearsing to you, gentlemen, the Town-Ho, Sperm Whaler of Nantucket, was cruising in your Pacific here, not very many days' sail westward from the eaves of this good Golden Inn. She was somewhere to the northward of the Line. One morning upon handling the pumps, according to daily usage, it was observed that she made more water in her hold than common. They supposed a sword-fish had stabbed her, gentlemen. But the captain , having some unusual reason for believing that rare good luck awaited him in those latitudes; and therefore being very averse to quit them, and the leak not being then considered at all dangerous, though, indeed, they could not find it after searching the hold as low down as was possible in rather heavy weather, the ship still continued her cruisings, the mariners working at the pumps at wide and easy intervals; but no good luck came; more days went by, and not only was the leak yet undiscovered, but it sensibly increased. So much so, that now taking some alarm, the captain, making all sail, stood away for the nearest harbor among the islands, there to have his hull hove out and repaired.

"Though no small passage was before her, yet, if the commonest chance favored, he did not at all fear that his ship would founder by the way, because his pumps were of the best, and being periodically relieved at them, those six-and-thirty men of his could easily keep the ship free; never mind if the leak should double on her. In truth, well nigh the whole of this passage being attended by very prosperous breezes, the Town-Ho had all but certainly arrived in perfect safety at her port without the occurrence of the least fatality, had it not been for the brutal overbearing of Radney, the mate, a Vine yarder, and the bitterly provoked vengeance of Steelkilt, a Lakeman and desperado from Buffalo."

"'Lakeman!—Buffalo! Pray, what is a Lakeman, and where is Buffalo?' said Don Sebastian, rising in his swinging mat of grass.

"On the eastern shore of our Lake Erie, Don; but—I crave your courtesy—may be, you shall soon hear further of all that Now, gentlemen, in square-sail brigs and three-masted ships well nigh as large and stout as any that ever sailed out of your old Callao to far Manilla; this Lakeman, in the land-locked heart of our America, has yet been nurtured by all those agrarian freebooting impressions popularly connected with the open ocean. For in their interflowing aggregate, those grand freshwater seas of ours,—Erie, and Ontario, and Huron, and Superior, and Michigan,—possess an ocean-like expansiveness, with many of the ocean's noblest traits; with many of its rimmed varieties of races and of climes. They contain round archipelagoes of romantic isles, even as the Polynesian waters do; in large part, are shored by two great contrasting nations, as the Atlantic is; they furnish long maritime approaches to our numerous territorial colonies from the East, dotted all round their banks; here and there are frowned upon by batteries, and by the goat-like craggy guns of lofty Mackinaw; they have heard the fleet thunderings of naval victories; at intervals, they yield their beaches to wild barbarians, whose red painted faces flash from out their peltry wigwams; for leagues and leagues are flanked by ancient and unentered forests, where the gaunt pines stand like serried lines of kings in Gothic genealogies; those same woods harboring wild Africa beasts of prey, and silken creatures whose exported furs give robes to Tartar Emperors; they mirror the paved capitals of Buffalo and Cleveland, as well as Winnebago villages; they float alike the full-rigged merchant ship, the armed cruiser of the State, the steamer, and the beech canoe; they are swept by Borean and dismasting blasts as direful as any that lash the salted wave; they know what shipwrecks are, for out of sight of land however inland, they have drowned full many a midnight ship with all its shrieking crew. Thus, gentlemen, though an inlander, Steelkilt was wild-ocean born, and wild-ocean nurtured; as much of an audacious mariner as any. And for Radney, though in his infancy he may have laid him down on the lone Nantucket beach, to nurse at his maternal sea; though in after life he had long followed our austere Atlantic and your contemplative Pacific; yet was he quite as vengeful and full of social quarrel as the backwoods seaman, fresh from the latitudes of buckhorn-handled Bowie-knives. Yet was this Nantucketer a man with some good-hearted traits; and this Lakeman, a mariner, who though a sort of devil indeed, might yet by inflexible firmness, only tempered by that common decency of human recognition which is the meanest slave's right; thus treated, this Steelkilt had long been retained harmless and docile. At all events, he had proved so thus far; but Radney was doomed and made mad, and Steelkilt—but, gentlemen, you shall hear.

"It was not more than a day or two at the furthest after pointing her prow for her island haven, that the Town-Ho's leak seemed again increasing, but only so as to require an hour or more at the pumps every day. You must know that in a settled and civilized ocean like our Atlantic, for example, some skippers think little of pumping their whole way across it; though of a still, sleepy night, should the officer of the deck happen to forget his duty in that respect, the probability would be that he and his shipmates would never again remember it, on account of all hands gently subsiding to the bottom. Nor in the solitary and savage seas far from you to the westward, gentlemen, is it altogether unusual for ships to keep clanging at their pump-handles in full chorus even for a voyage of considerable length; that is, if it lie along a tolerably accessible coast, or if any other reasonable retreat is afforded them. It is only when a leaky vessel is in some very out of the way part of those waters, some really landless latitude, that her captain begins to feel a little anxious.

"Much this way had it been with the Town-Ho; so when her leak was found gaining once more, there was in truth some small concern manifested by several of her company, especially by Radney the mate. He commanded the upper sails to be well hoisted, sheeted home anew, and every way expanded to the breeze. Now this Radney, I suppose, was as little of a coward, and as little inclined to any sort of nervous apprehensiveness touching his own person as any fearless, unthinking creature on land or on sea that you can conveniently imagine, gentlemen. Therefore when he betrayed this solicitude about the safety of the ship, some of the seamen declared that it was only on account of his being a part owner in her. So when they were working that evening at the pumps, there was on this head no small gamesomeness slyly going on among them, as they stood with their feet continually overflowed by the rippling clear water; clear as any mountain spring, gentlemen, that bubbling from the pumps ran across the deck, and poured itself out in steady spouts at the lee scupper-holes.

"Now, as you well know, it is not seldom the case in this con-ventional world of ours—watery or otherwise—that when a person placed in command over his fellow-men finds one of them to be very significantly his superior in general pride of manhood, straightway against that man he conceives an unconquerable dislike and bitterness; and if he have a chance he will pull down and pulverize that subaltern's tower, and make a little heap of dust of it. Be this conceit of mine as it may, gentlemen, at all events Steelkilt was a tall and noble animal with a head like a Roman, and a flowing golden beard like the tasseled housings of your last viceroy's snorting charger; and a brain, and a heart, and a soul in him, gentlemen, which had made Steelkilt Char-lemagne, had he been born son to Charlemagne's father. But Radney, the mate, was ugly as a mule; yet as hardy, as stubborn, as malicious. He did not love Steelkilt, and Steelkilt knew it.

'Espying the mate drawing near as he was toiling at the pump with the rest, the Lakeman affected not to notice him, but unawed, went on with his gay banterings.

"'Aye, aye, my merry lads, it's a lively leak this; hold a cannikin, one of ye, and let's have a taste. By the Lord, it's worth bottling! I tell ye what, men, old Rad's investment must go for it! He had best cut away his part of the hull and tow it home. The fact is, boys, that sword-fish only began the job; he's come back again with a gang of ship-carpenters, saw-fish, and file-fish, and what not; and the whole posse of 'em are now hard at work cutting and slashing at the bottom; making improvements, I suppose. If old Rad were here now, I'd tell him to jump overboard and scatter 'em. They're playing the devil with his estate, I can tell him. But he's a simple old soul,—Rad—and a beauty too. Boys, they say the rest of his property is invested in looking-glasses. I wonder if he'd give a poor devil like me the model of his nose.'

"'Damn your eyes! What's that pump stopping for?' roared Radney, pretending not to have heard the sailors' talk. 'Thunder away at it!'

"'Aye, aye, sir,' said Steelkilt, merry as a cricket. 'Lively, boys, lively, now!' And with that the pump clanged like fifty fire engines; the men tossed their hats off to it, and ere long that peculiar gasping of the lungs was heard which denotes the fullest tension of life's utmost energies.

"Quitting the pump at last, with the rest of his band, the Lakeman went forward all panting, and sat himself down on the windlass; his face fiery red, his eyes bloodshot, and wiping the profuse sweat from his brow. Now what cozening fiend it was, gentlemen, that possessed Radney to meddle with such a man in that corporeally exasperated state, I know not; but so it happened. Intolerably striding along the deck, the mate commanded him to get a broom and sweep down the planks, and also a shovel, and remove some offensive matters consequent upon allowing a pig to run at large.

"Now, gentlemen, sweeping a ship's deck at sea is a piece of household work which in all times but raging gales is regularly attended to every evening; it has been known to be done in the case of ships actually foundering at the time. Such, gentlemen, is the inflexibility of sea-usages and the instinctive love of neatness in seamen; some of whom would not willingly drown without first washing their faces. But in all vessels this broom business is the prescriptive province of the boys, if boys there be aboard. Besides, it was the stronger men in the Town-Ho that had been divided into gangs, taking turns at the pumps; and being the most athletic seaman of them all, Steelkilt had been regularly assigned captain of one of the gangs; consequently he should have been freed from any trivial business not connected with truly nautical duties, such being the case with his comrades. I mention all these particulars so that you may understand exactly how this affair stood between the two men.

"But there was more than this: the order about the shovel was almost as plainly meant to sting and insult Steelkilt, as though Radney had spat in his face. Any man who has gone sailor in a whale-ship will understand this; and all this and doubtless much more, the Lakeman fully comprehended when the mate uttered his command. But as he sat still for a moment, and as he steadfastly looked into the mate's malignant eye and perceived the stacks of powder-casks heaped up in him and the slow-match silently burning along towards them; as he instinctively saw all this, that strange forbearance and unwillingness to stir up the deeper passionateness in any already ireful being—a re-pugnance most felt, when felt at all, by really valiant men even when aggrieved—this nameless phantom feeling, gentlemen, stole over Steelkilt.

"Therefore, in his ordinary tone, only a little broken by the bodily exhaustion he was temporarily in, he answered his saying that sweeping the deck was not his business, and he would not do it. And then, without at all alluding to the shovel, he pointed to three lads as the customary sweepers; who, not being billeted at the pumps, had done little or nothing all day. To this, Radney replied with an oath, in a most domineering and outrageous manner unconditionally reiterating his command; meanwhile advancing upon the still seated Lakeman, with an uplifted cooper's club hammer which he had snatched from a cask nearby.

"Heated and irritated as he was by his spasmodic toil at the pumps, for all his first nameless feeling of forbearance the sweating Steelkilt could but ill brook this bearing in the mate; but somehow still smothering the conflagration within him, without speaking he remained doggedly rooted to his seat, till at last the incensed Radney shook the hammer within a few inches of his face, furiously commanding him to do his bidding.

"Steelkilt rose, and slowly retreating round the windlass, steadily followed by the mate with his menacing hammer, deliberately repeated his intention not to obey. Seeing, however, that his forbearance had not the slightest effect, by an awful and unspeakable intimation with his twisted hand he warned off the foolish and infatuated man; but it was to no purpose. And in this way the two went once slowly round the windlass; when, resolved at last no longer to retreat, bethinking him that he had now forborne as much as comported with his humor, the Lakeman paused on the hatches and thus spoke to the officer:

"'Mr. Radney, I will not obey you. Take that hammer away, or look to yourself.' But the predestinated mate coming still closer to him, where the Lakeman stood fixed, now shook the heavy hammer within an inch of his teeth; meanwhile repeating a string of insufferable maledictions. Retreating not the thousandth part of an inch: stabbing him in the eye with the unflinching poniard of his glance, Steelkilt, clenching his right hand behind him and creepingly drawing it back, told his persecutor that if the hammer but grazed his cheek he (Steelkilt) would murder him. But, gentlemen, the fool had been branded for the slaughter by the gods. Immediately the hammer touched the cheek; the next instant the lower jaw of the mate was stove in his head; he fell on the hatch spouting blood like a whale.

"Ere the cry could go aft Steelkilt was shaking one of the back-stays leading far aloft to where two of his comrades were standing their mastheads. They were both Canallers.

"'Canallers!' cried Don Pedro. 'We have seen many whale ships in our harbors, but never heard of your Canallers Pardon: who and what are they?'

"'Canallers, Don, are the boatmen belonging to our grand Erie Canal. You must have heard of it.'

"'Nay, Senor; hereabouts in this dull, warm, most lazy, and hereditary land, we know but little of your vigorous North.'

"'Aye? Well then, Don, refill my cup. Your chicha's very fine; and ere proceeding further I will tell ye what our Canallers are; for such information may throw side-light upon my story.'

"For three hundred and sixty miles, gentlemen, through the entire breadth of the state of New York; through numerous populous cities and most thriving villages; through long, dismal, uninhabited swamps, and affluent, cultivated fields, unrivalled for fertility; by billiard-room and bar-room; through the holy-of-holies of great forests; on Roman arches over Indian rivers; through sun and shade; by happy hearts or broken; through all the wide contrasting scenery of those noble Mohawk counties, and especially, by rows of snow-white chapels, whose spires stand almost like milestones, flows one continual stream of Venetianly corrupt and often lawless life. There's your true Ashantee, gentlemen; there howl your pagans; where you ever find them, next door to you; under the long-flung shadow, and the snug patronizing lee of churches. For by some curious fatality, as it is often noted of your metropolitan freebooters that they ever encamp around the halls of justice, so sinners, gentlemen, most abound in holiest vicinities.

"'Is that a friar passing?' said Don Pedro, looking downwards into the crowded plazza, with humorous concern.

"'Well for our northern friend, Dame Isabella's Inquisition wanes in Lima,' laughed Don Sebastian. 'Proceed, Senor.'

"'A moment! Pardon!' cried another of the company. "'In the name of all us Limeese, I but desire to express to you, sir sailor, that we have by no means overlooked your delicacy in not substituting present Lima for distant Venice in your corrupt comparison. Oh! do not bow and look surprised; you know the proverb all along this coast—"Corrupt as Lima." It but bears out your saying, too; churches more plentiful than billiard-tables, and forever open—and "Corrupt as Lima." So, too, Venice; I have been there; the city of the blessed evangelist, St. Mark!—St. Dominic, purge it! Your cup! Thanks: here I refill; now, you pour out again.'

"Freely depicted in his own vocation, gentlemen, the Canaller would make a fine dramatic hero, so abundantly and picturesquely wicked is he. Like Mark Antony, for days and days along his green-turfed, flowery Nile, he indolently floats, openly toying with his red-cheeked Cleopatra, ripening his apricot thigh upon the sunny deck. But ashore, all this effeminacy is dashed. The brigandish guise which the Canaller so proudly sports; his slouched and gaily-ribboned hat betoken his grand features. A terror to the smiling innocence of the villages through which he floats; his swart visage and bold swagger are not unshunned in cities. Once a vagabond on his own canal, I have received good turns from one of these Canallers; I thank him heartily; would fain be not ungrateful; but it is often one of the prime redeeming qualities of your man of violence, that at times he has as stiff an arm to back a poor stranger in a strait, as to plunder a wealthy one. I sum, gentlemen, what the wildness of this canal life is, is emphatically evinced by this; that our wild whale-fishery contains so many of its most finished graduates, and that scarce any race of mankind, except Sydney men, are so much distrusted by our whaling captains. Nor does it at all diminish the curiousness of this matter, that to many thousands of our rural boys and young men born along its line, the probationary life of the Grand Canal furnishes the sole transition between quietly reaping in a Christian corn-field, and recklessly ploughing the waters of the most barbaric seas."

"'I see! I see!' impetuously exclaimed Don Pedro, spilling his chicha upon his silvery ruffles. 'No need to travel! The world's one Lima. I had thought, now, that at your temperate North the generations were cold and holy as the hills.—But the story.'

"I left off, gentlemen, where the Lakeman shook the backstay. Hardly had he done so, when he was surrounded by the three junior mates and the four harpooneers, who all crowded him to the deck. But sliding down the ropes like baleful comets, the two Canallers rushed into the uproar, and sought to drag their man out of it towards the forecastle. Others of the sailors joined with them in this attempt, and a twisted turmoil ensued; while standing out of harm's way, the valiant captain danced up and down with a whale-pike, calling upon his officers to manhandle that atrocious scoundrel, and smoke him along to the quarter-deck. At intervals, he ran close up to the revolving border of the confusion, and prying into the heart of it with his pike, sought to pick out the object of his resentment. But Steelkilt and his desperadoes were too much for them all; they succeeded in gaining the forecastle deck, where, hastily slewing about three or four large casks in a line with the windlass, these sea-Parisians entrenched themselves behind the barricade.

"'Come out of that, ye pirates!' roared the captain, now menacing them with a pistol in each hand, just brought to him by the steward. 'Come out of that, ye cut-throats!'

"Steelkilt leaped on the barricade, and striding up and down there, defied the worst the pistols could do; but gave the captain to understand distinctly, that his (Steelkilt's) death would be the signal for a murderous mutiny on the part of all hands.

Fearing in his heart lest this might prove but too true, the captain a little desisted, but still commanded the insurgents instantly to return to their duty.

"'Will you promise not to touch us, if we do?' demanded their ringleader.

"'Turn to! turn to!—I make no promise;—to your duty! Do you want to sink the ship, by knocking off at a time like this? Turn to!' and he once more raised a pistol.

"'Sink the ship?' cried Steelkilt. 'Aye, let her sink. Not a man of us turns to, unless you swear not to raise a rope-yarn against us. What say ye, men?' turning to his comrades. A fierce cheer was their response.

"The Lakeman now patrolled the barricade, all the while keeping his eye on the Captain, and jerking out such sentences as these:—'It's not our fault; we didn't want it; I told him to take his hammer away; it was boy's business; he might have known me before this; I told him not to prick the buffalo; I believe I have broken a finger here against his cursed jaw; ain't those mincing knives down in the forecastle there, men? look to those handspikes, my hearties. Captain, by God, look to yourself; say the word; don't be a fool; forget it all; we are ready to turn to; treat us decently, and we're your men; but we won't be flogged.'

"'Turn to! I make no promises, turn to, I say!'

"'Look ye, now,' cried the Lakeman, flinging out his arm towards him, 'there are a few of us here (and I am one of them) who have shipped for the cruise, d'ye see; now as you well know, sir, we can claim our discharge as soon as the anchor is down; so we don't want a row; it's not our interest; we want to be peaceable; we are ready to work, but we won't be flogged.'

"'Turn to!' roared the Captain.

"'Steelkilt glanced round him a moment, and then said:—'I tell you what it is now, Captain, rather than kill ye, and be hung for such a shabby rascal, we won't lift a hand against ye unless ye attack us; but till you say the word about not flogging us, we don't do a hand's turn.'

"'Down into the forecastle then, down with ye, I'll keep ye there till ye're sick of it. Down ye go.'

"'Shall we?' cried the ringleader to his men. Most of them were against it; but at length, in obedience to Steelkilt, they preceded him down into their dark den, growlingly disappearing, like bears into a cave.

"As the Lakeman's bare head was just level with the planks, the Captain and his posse leaped the barricade, and rapidly drawing over the slide of the scuttle, planted their group of hands upon it, and loudly called for the steward to bring the heavy brass padlock belonging to the companion-way. Then opening the slide a little, the Captain whispered something down the crack, closed it, and turned the key upon them—ten in number—leaving on deck some twenty or more, who thus far had remained neutral.

"All night a wide-awake watch was kept by all the officers, forward and aft, especially about the forecastle scuttle and fore hatchway, at which last place it was feared the insurgents might emerge, after breaking through the bulkhead below. But the hours of darkness passed in peace; the men who still remained at their duty toiling hard at the pumps, whose clinking and clanking at intervals through the dreary night dismally resounded through the ship.

"At sunrise the Captain went forward, and knocking on the deck, summoned the prisoners to work; but with a yell they refused. Water was then lowered down to them, and couple of handfuls of biscuit were tossed after it; when again turning the key upon them and pocketing it, the Captain returned to the quarter-deck. Twice every day for three days this was repeated; but on the fourth morning a confused wrangling, and then a scuffling was heard, as the customary summons was delivered; and suddenly four men burst up from the forecastle saying they were ready to turn to. The fetid closeness of the air, and a famishing diet, united perhaps to some fears of ultimate retribution, has constrained them to surrender at discretion. Emboldened by this, the Captain reiterated his demand to the rest, but Steelkilt shouted up to him a terrific hint to stop his babbling and betake himself where he belonged. On the fifth morning three others of the mutineers bolted up into the air from the desperate arms below that sought to restrain them. Only three were left.

"'Better turn to, now?' said the Captain with a heartless jeer.

"'Shut us up again, will ye!' cried Steelkilt.

"'Oh! certainly,' said the Captain, and the key clicked.

"It was at this point, gentlemen, that enraged by the defection of seven of his former associates, and stung by the mocking voice that had last hailed him, and maddened by his long entombment in a place as black as the bowels of despair; it was then that Steelkilt proposed to the two Canallers, thus far apparently of one mind with him, to burst out of their hole at the next summoning of the garrison; and armed with their keen mincing knives (long, crescentic, heavy implements with a handle at each end) run a muck from the bowsprit to the taffrail; and if by any devilishness of desperation possible, seize the ship. For himself, he would do this, he said, whether they joined him or not. That was the last night he should spend in that den. But the scheme met with no opposition on the part of the other two; they swore they were ready for that, or for any other mad thing, for anything in short but a surrender. And what was more, they each insisted upon being the first man on deck, when the time to make the rush should come.

But to this their leader as fiercely objected, reserving that priority for himself; particularly as his two comrades would not yield, the one to the other, in the matter; and both of them could not be first, for the ladder would but admit one man at a time. And here, gentlemen, the foul play of these miscreants must come out.

"Upon hearing the frantic project of their leader, each in his own separate soul had suddenly lighted, it would seem, upon the same piece of treachery, namely: to be foremost in breaking out, in order to be the first of the three, though the last of the ten, to surrender; and thereby secure whatever small chance of pardon such conduct might merit. But when Steelkilt made known his determination still to lead them to the last, they in some way, by some subtle chemistry of villany, mixed their before secret treacheries together; and when their leader fell into a doze, verbally opened their souls to each other in three sentences; and bound the sleeper with cords, and gagged him with cords; and shrieked out for the Captain at midnight.

"Thinking murder at hand, and smelling in the dark for the blood, he and all his armed mates and harpooneers rushed for the forecastle. In a few minutes the scuttle was opened, and, bound hand and foot, the still struggling ringleader was shoved up into the air by his perfidious allies, who at once claimed the honor of securing a man who had been fully ripe for murder. But all these were collared, and dragged along the deck like dead cattle; and, side by side, were seized up into the mizzen rigging, like three quarters of meat, and there they hung till morning. 'Damn ye,' cried the Captain, pacing to and fro before them 'the vultures would not touch ye, ye villains!'

"At sunrise he summoned all hands; and separating those who had rebelled from those who had taken no part in the mutiny, he told the former that he had a good mind to flog them all round—thought, upon the whole, he would do so—he ought to—justice demanded it; but for the present, considering their timely surrender, he would let them go with a reprimand, which he accordingly administered in the vernacular.

"'But as for you, ye carrion rogues,' turning to the three men in the rigging—'for you, I mean to mince ye up for the try-pots; and, seizing a rope, he applied it with all his might to the backs of the two traitors, till they yelled no more, but lifelessly hung their heads sideways, as the two crucified thieves are drawn.

"'My wrist is sprained with ye!' he cried, at last; 'but there is still rope enough left for you, my fine bantam, that wouldn't give up. Take that gag from his mouth, and let us hear what he can say for himself.'

"For a moment the exhausted mutineer made a tremulous motion of his cramped jaws, and then painfully twisting round his head, said in a sort of hiss, 'What I say is this—and mind it well—if you flog me, I murder you!'

"'Say ye so? then see how ye frighten me'—and the Captain drew off with the rope to strike.

"'Best not,' hissed the Lakeman.

"'Best I must,'—and the rope was once more drawn back for the stroke.

"Steelkit here hissed out something, inaudible to all but the Captain; who, to the amazement of all hands, started back, paced the deck rapidly two or three times, and then suddenly throwing down his rope, said, 'I won't do it—let him go—cut him down: d'ye hear?'

"But as the junior mates were hurrying to execute the order a pale man, with a bandaged head, arrested them—Radney the chief mate. Ever since the blow, he had lain in his berth; but that morning, hearing the tumult on the deck, he had crept out, and thus far had watched the whole scene. Such was the state of his mouth, that he could hardly speak; but mumbling something about his being willing and able to do what the captain dared not attempt, he snatched the rope and advanced to his pinioned foe.

"'You are a coward!' hissed the Lakeman.

"'So I am, but take that.' The mate was in the very act of striking, when another hiss stayed his uplifted arm. He paused: and then pausing no more, made good his word, spite of Steelkilt's threat, whatever that might have been. The three men were then cut down, all hands were turned to, and, sullenly worked by the moody seamen, the iron pumps clanged as before.

"Just after dark that day, when one watch had retired below, a clamor was heard in the forecastle; and the two trembling traitors running up, besieged the cabin door, saying they durst not consort with the crew. Entreaties, cuffs, and kicks could not drive them back, so at their own instance they were put down in the ship's run for salvation. Still, no sign of mutiny reappeared among the rest. On the contrary, it seemed, that mainly at Steelkilt's instigation, they had resolved to maintain the strictest peacefulness, obey all orders to the last, and, when the ship reached port, desert her in a body. But in order to insure the speediest end to the voyage, they all agreed to another thing—namely, not to sing out for whales, in case any should be discovered. For, spite of her leak, and spite of all her other perils, the Town-Ho still maintained her mastheads, and her captain was just as willing to lower for a fish that moment, as on the day his craft first struck the cruising ground; and Radney the mate was quite as ready to change his berth for a boat, and with his bandaged mouth seek to gag in death the vital jaw of the whale.

"But though the Lakeman had induced the seamen to adopt this sort of passiveness in their conduct, he kept his own counsel (at least till all was over) concerning his own proper and private revenge upon the man who had stung him in the ventricles of his heart. He was in Radney the chief mate's watch; and as if the infatuated man sought to run more than half way to meet his doom, after the scene at the rigging, he insisted, against the express counsel of the captain, upon resuming the head of his watch at night. Upon this, and one or two other circumstances, Steelkilt systematically built the plan of his revenge.

"During the night, Radney had an unseamanlike way of sitting on the bulwarks of the quarter-deck, and leaning his arm upon the gunwale of the boat which was hoisted up there, a little above the ship's side. In this attitude, it was well known, he sometimes dozed. There was a considerable vacancy between the boat and the ship, and down between this was the sea. Steelkilt calculated his time, and found that his next trick at the helm would come round at two o'clock, in the morning of the third day from that in which he had been betrayed. At his leisure, he employed the interval in braiding something very carefully in his watches below.

"'What are you making there?' said a shipmate.

"'What do you think? What does it look like?'

"'Like a lanyard for your bag; but it's an odd one, seems to me.'

"'Yes, rather oddish,' said the Lakeman, holding it at arm's length before him; 'but I think it will answer. Shipmate, I haven't enough twine—have you any?'

"But there was none in the forecastle.

"'Then I must get some from old Rad;' and he rose to go aft.

"'You don't mean to go a begging to him!' said a sailor.

"'Why not? Do you think he won't do me a turn, when it's to help himself in the end, shipmate?' and going to the mate, he looked at him quietly, and asked him for some twine to mend his hammock. It was given him—neither twine nor lanyard were seen again; but the next night an iron ball, closely netted, partly rolled from the pocket of the Lakeman's monkey jacket, as he was tucking the coat into his hammock for a pillow. Twenty-four hours after his trick at the silent helm—night to the man who was apt to doze over the grave already dug by the seaman's hand—that fatal hour was then to come; and in the fore-ordaining soul of Steelkilt, the mate was already stark and stretched as a corpse, with his forehead crushed in.

"But, gentlemen, a fool saved the would-be murderer from the bloody deed he had planned. Yet complete revenge he had, and without being the avenger. For by a mysterious fatality, Heaven itself seemed to step in to take out of his hands into its own the damning thing he would have done.

"It was just between daybreak and sunrise of the morning of the second day, when they were washing down the decks, that a stupid Teneriffe man, drawing water in the main-chains, all at once shouted out, 'There she rolls! there she rolls!' Jesu, what a whale! It was Moby Dick.

"'Moby Dick!' cried Don Sebastian; 'St. Dominic! Sir sailor, but do whales have christenings? Whom call you Moby Dick?'

"'A very white, and famous, and most deadly immortal monster, Don;—but that would be too long a story.'

"'How? how?' cried all the young Spaniards, crowding.

"'Nay, Dons, Dons—nay, nay! I cannot rehearse that now.

Let me get more into the air, Sirs.'

"'The chicha! the chicha!' cried Don Pedro; 'our vigorous friend looks faint;—fill up his empty glass!'

"No need, gentlemen; one moment, and I proceed.—Now, gentlemen, so suddenly perceiving the snowy whale within fifty yards of the ship—forgetful of the compact among the crew—in the excitement of the moment, the Teneriffe man had instinctively and involuntarily lifted his voice for the monster, though for some little time past it had been plainly beheld from the three sullen mastheads. All was now a phrensy. 'The White Whale—the White Whale!' was the cry from captain, mates, and harpooneers, who, undeterred by fearful rumors, were all anxious to capture so famous and precious a fish; while the dogged crew eyed askance, and with curses, the appalling beauty of the vast milky mass, that lit up by a horizontal spangling sun, shifted and glistened like a living opal in the blue morning sea. Gentlemen, a strange fatality pervades the whole career of these events, as if verily mapped out before the world itself was charted. The mutineer was the bowsman of the mate, and when fast to a fish, it was his duty to sit next him, while Radney stood up with his lance in the prow, and haul in or slacken the line, at the word of command. Moreover, when the four boats were lowered, the mate's got the start; and none howled more fiercely with delight than did Steelkilt, as he strained at his oar. After a stiff pull, their harpooneer got fast, and, spear in hand, Radney sprang to the bow. He was always a furious man it seems, in a boat. And now his bandaged cry was to beach him on the whale's topmost back. Nothing loath, his bowsman hauled him up and up, through a blinding foam that blent two whitenesses together; till of a sudden the boat struck as against a sunken ledge, and keeling over, spilled out the standing mate. That instant, as he fell on the whale's slippery back, the boat righted, and was dashed aside by the swell, while Radney was tossed over into the sea, on the other flank of the whale. He struck out through the spray, and for an instant, was dimly seen through that veil, wildly seeking to remove himself from the eye of Moby Dick. But the whale rushed round in a sudden maelstrom; seized the swimmer between his jaws;

and rearing high up with him, plunged headlong again, and went down.

"Meantime, at the first tap of the boat's bottom, the Lakeman had slackened the line, so as to drop astern from the whirlpool; calmly looking on, he thought his own thoughts. But a sudden, terrific, downward jerking of the boat, quickly brought his knife to the line. He cut it; and the whale was free. But, at some distance, Moby Dick rose again, with some tatters of Radney's red woollen shirt, caught in the teeth that had destroyed him. All four boats gave chase again; but the whale eluded them, and finally wholly disappeared.

"In good time, the Town-Ho reached her port—a savage, solitary place—where no civilized creature resided. There, headed by the Lakeman, all but five or six of the foremast-men deliberately deserted among the palms; eventually, as it turned out, seizing a large double war-canoe of the savages, and setting sail for some other harbor.

"The ship's company being reduced to but a handful, the captain called upon the Islanders to assist him in the laborious business of heaving down the ship to stop the leak. But to such unresting vigilance over their dangerous allies was this small band of whites necessitated, both by night and by day, and so extreme was the hard work they underwent, that upon the vessel being ready again for sea, they were in such a weakened condition that the captain durst not put off with them in so heavy a vessel. After taking counsel with his officers, he anchored the ship as far off shore as possible; loaded and ran out his two cannon from the bows; stacked his muskets on the poop; and warning the Islanders not to approach the ship at their peril, took one man with him, and setting the sail of his best whale-boat, steered straight before the wind for Tahiti, five hundred miles distant, to procure a reinforcement to his crew.

"On the fourth day of the sail, a large canoe was descried, which seemed to have touched at a low isle of corals. He steered away from it; but the savage craft bore down on him; and soon the voice of Steelkilt hailed him to heave to, or he would run him under water. The captain presented a pistol. With one foot on each prow of the yoked war-canoes, the Lakeman laughed him to scorn; assuring him that if the pistol so much as clicked in the lock, he would bury him in bubbles and foam.

"'What do you want of me?" cried the captain.

"'Where are you bound? and for what are you bound?' demanded Steelkilt; 'no lies.'

"'I am bound to Tahiti for more men.'

"'Very good. Let me board you a moment—I come in peace.'

With that he leaped from the canoe, swam to the boat; and climbing the gunwhale, stood face to face with the captain.

"'Cross your arms, sir; throw back your head. Now, repeat after me. As soon as Steelkilt leaves me, I swear to beach this boat on yonder island, and remain there six days. If I do not, may lightnings strike me!'

"'A pretty scholar,' laughed the Lakeman. 'Adios, Senor!' and leaping into the sea, he swam back to his comrades.

"'Watching the boat till it was fairly beached, and drawn up to the roots of the cocoa-nut trees, Steelkilt made sail again, and in due time arrived at Tahiti, his own place of destination. There, luck befriended him; two ships were about to sail for France, and were providentially in want of precisely that number of men which the sailor headed. They embarked; and so for ever got the start of their former captain, had he been at all minded to work them legal retribution.

"Some ten days after the French ships sailed, the whale-boat arrived, and the captain was forced to enlist some of the more civilized Tahitians, who had been somewhat used to the sea. Chartering a small native schooner, he returned with them to his vessel; and finding all right there, again resumed his cruisings.

"Where Steelkilt now is, gentlemen, none know; but upon the island of Nantucket, the widow of Radney still turns to the sea which refuses to give up its dead; still in dreams sees the awful white whale that destroyed him.

*   *   *

"'Are you through?' said Don Sebastian, quietly.

"'I am, Don.'

"'Then I entreat you, tell me if to the best of your own convictions, this your story is in substance really true? It is so passing wonderful! Did you get it from an unquestionable source? Bear with me if I seem to press.'

"'Also bear with all of us, sir sailor; for we all join in Don Sebastian's suit,' cried the company, with exceeding interest.

"'Is there a copy of the Holy Evangelists in the Golden Inn, gentlemen?'

"'Nay,' said Don Sebastian; 'but I know a worthy priest near by, who will quickly procure one for me. I go for it; but are you well advised? This may grow too serious.'

"'Will you be so good as to bring the priest also, Don?'

"'Though there are no Auto-da-Fés in Lima now,' said one of the company to another; 'I fear our sailor friend runs risk of the archiepiscopacy. Let us withdraw more out of the moonlight. I see no need of this.'

"'Excuse me for running after you, Don Sebastian; but may I also beg that you will be particular in procuring the largest sized Evangelists you can.'

*   *   *

"'This is the priest, he brings you the Evangelists,' said Don Sebastian, gravely, returning with a tall and solemn figure.

"'Let me remove my hat. Now, venerable priest, further into the light, and hold the Holy Book before me that I may touch it.'

"'So help me Heaven, and on my honor the story I have told ye, gentlemen, is in substance and its great items, true. I know it to be true; it happened on this ball; I trod the ship; I knew the crew; I have seen and talked with Steelkilt since the death of Radney.'"

CHAPTER 10

"THE PIRATE CREW SET SAIL" FROM *THE ADVENTURES OF TOM SAWYER*

by Mark Twain

Drawing from the 1876 edition of *The Adventures of Tom Sawyer.*

Tom's mind was made up now. He was gloomy and desperate. He was a forsaken, friendless boy, he said; nobody loved him; when they found out what they had driven him to, perhaps they would be sorry; he had tried to do right and get along, but they would not let him, since nothing would do them but to be rid of him, let it be so; and let them blame him for the consequences—why shouldn't they? What right had the friendless to complain? Yes, they had forced him to it at last: he would lead a life of crime. There was no choice.

By this time he was far down Meadow Lane, and the bell for school to "take up" tinkled faintly upon his ear. He sobbed, now, to think he should never, never hear that old familiar sound any more—it was very hard, but it was forced on him; since he was driven out into the cold world, he must submit—but he forgave them. Then the sobs came thick and fast.

Just at this point he met his soul's sworn comrade, Joe Harper—hard-eyed, and with evidently a great and dismal purpose in his heart. Plainly here were "two souls with but a single thought." Tom, wiping his eyes with his sleeve, began to blubber out something about a resolution to escape from hard usage and lack of sympathy at home by roaming abroad into the great world never to return; and ended by hoping that Joe would not forget him.

But it transpired that this was a request which Joe had just been going to make to Tom, and had come to hunt him up for that purpose. His mother had whipped him for drinking some cream which he had never tasted and knew nothing about; it was plain that she was tired of him and wished him to go; if she felt that way, there was nothing for him to do but succumb; he hoped she would be happy, and never regret having driven her poor boy out into the unfeeling world to suffer and die.

As the two boys walked sorrowing along, they made a new compact to stand by each other and be brothers and never be separate till death relieved them of their troubles. Then they began to lay their plans. Joe was for being a hermit, and living on crusts in a remote cave, and dying, sometime, of cold and want and grief; but after listening to Tom, he conceded that there were some conspicuous advantages about a life of crime, and so he consented to be a pirate.

Three miles below St. Petersburg, at a point where the Mississippi River was a trifle over a mile wide, there was a long, narrow, wooded island, with a shallow bar at the head of it, and this offered well as a rendezvous. It was not inhabited; it lay far over toward the farther shore, abreast a dense and almost wholly unpeopled forest. So Jackson's Island was chosen. Who were to be the subjects of their piracies was a matter that did not occur to them. Then they hunted up Huckleberry Finn, and he joined them promptly, for all careers were one to him; he was indifferent. They presently separated to meet at a lonely spot on the riverbank two miles above the village at the favorite hour—which was midnight. There was a small log raft there which they meant to capture. Each would bring hooks and lines, and such provisions as he could steal in the most dark and mysterious way—as became outlaws. And before the afternoon was done they had all managed to enjoy the sweet glory of spreading the fact that pretty soon the town would "hear something." All who got this vague hint were cautioned to "be mum and wait."

About midnight Tom arrived with a boiled ham and a few trifles, and stopped in a dense undergrowth on a small bluff overlooking the meeting place. It was starlight, and very still. The mighty river lay like an ocean at rest. Tom listened a moment, but no sound disturbed the quiet. Then he gave a low, distinct whistle. It was answered from under the bluff. Tom whistled twice more; these signals were answered in the same way. Then a guarded voice said:

"Who goes there?"

"Tom Sawyer, the Black Avenger of the Spanish Main. Name your names."

"Huck Finn the Red-Handed, and Joe Harper the Terror of the Seas." Tom had furnished these titles, from his favorite literature.

"'Tis well. Give the countersign."

Two hoarse whispers delivered the same awful word simultaneous to the brooding night;

"BLOOD!"

Then Tom tumbled his ham over the bluff and let himself down after it, tearing both skin and clothes to some extent in the effort. There was an easy, comfortable path along the shore under the bluff, but it lacked the advantages of difficulty and danger so valued by a pirate.

The Terror of the Seas had brought a side of bacon, and had about worn himself out with getting it there. Finn the Red-Handed had stolen a skillet and a quantity of half-cured leaf tobacco, and had also brought a few corncobs to make pipes with. But none of the pirates smoked or "chewed" but himself. The Black Avenger of the Spanish Main said it would never do to start without some fire. That was a

Boy on a Raft by Winslow Homer, 1879. Courtesy of the National Gallery of Art.

wise thought; matches were hardly known there in that day. They saw a fire smoldering upon a great raft a hundred yards above, and they went stealthily thither and helped themselves to a chunk. They made an imposing adventure of it, saying, "Hist!" every now and then, and suddenly halting with finger on lip; moving with hands on imaginary dagger hilts; and giving orders in dismal whispers that if "the foe" stirred to "let him have it to the hilt," because "dead men tell no tales." They knew well enough that the raftsmen were all down at the village laying in stores or having a spree, but still that was no excuse for their conducting this thing in an unpiratical way.

They shoved off, presently, Tom in command, Huck at the after oar and Joe at the forward. Tom stood amidships, gloomy browed, and with folded arms, and gave his orders in a low, stern whisper:

"Luff, and bring her to the wind!"

"Aye-aye, sir!"

"Steady, steady-y-y-y!"

"Steady it is, sir!"

"Let her go off a point!"

"Point it is, sir!"

As the boys steadily and monotonously drove the raft toward midstream it was no doubt understood that these orders were given only for "style," and were not intended to mean anything in particular.

"What sail's she carrying?"

"Courses, tops'ls, and flying jib, sir."

"Send the r'yals up! Lay out aloft, there, half a dozen of ye—fore-topmaststuns'! Lively, now!"

"Aye-aye, sir!"

"Shake out that maintogalans'! Sheets and braces! Now, my hearties!"

"Aye-aye, sir!"

"Hellum-a-lee—hard aport! Stand by to meet her when she comes! Port, port! Now, men! With a will! Steady-y-y-y!"

"Steady it is, sir!"

The raft drew beyond the middle of the river; the boys pointed her head right, and then lay on their oars. The river was not high, so there was not more than a two- or three-mile current. Hardly a word was said during the next three-quarters of an hour. Now the raft was passing before the distant town. Two or three glimmering lights showed where it lay, peacefully sleeping, unconscious of the tremendous event that was happening. The Black Avenger stood still with folded arms, "looking his last" upon the scene of his former joys and his later sufferings, and wishing "she" could see him now, abroad on the wild sea, facing peril and death with dauntless heart, going to his doom with a grim smile on his lips. It was but a small strain on his imagination to remove Jackson's Island beyond eyeshot of the village, and so he "looked his last" with a broken and satisfied heart.

The other pirates were looking their last, too; and they all looked so long that they came near letting the current drift them out of the range of the island. But they discovered the danger in time, and made shift to avert it. About two o'clock in the morning the raft grounded on the bar two hundred yards above the head of the island, and they waded back and forth until they had landed their freight. Part of the little raft's belongings consisted of an old sail, and this they spread over a nook in the bushes for a tent to shelter their provisions; but they themselves would sleep in the open air in good weather, as became outlaws.

They built a fire against the side of a great log twenty or thirty steps within the somber depths of the forest, and then cooked some bacon in the frying pan for supper, and used up half of the corn "pone" stock they had brought. It seemed glorious sport to be feasting in that wild free way in the virgin forest of an unexplored and uninhabited island, far from the haunts of men, and they said they never would return to civilization. The climbing fire lit up their faces and threw its ruddy glare upon the pillared tree trunks of their forest temple, and upon the varnished foliage and festooning vines.

When the last crisp slice of bacon was gone and the last allowance of corn pone devoured, the boys stretched themselves out on the grass, filled with contentment. They could have found a cooler place, but they would not deny themselves such a romantic feature as the roasting campfire.

"Ain't it gay?" said Joe.

"It's nuts!" said Tom. "What would the boys say if they could see us?

"Say? Well, they'd just die to be here—hey, Hucky!"

"I reckon so," said Huckleberry; "anyways, I'm suited. I don't want nothing better'n this. I don't ever get enough to eat, gen'ally—and here they can't come and pick at a feller and bullyrag him so."

"It's just the life for me," said Tom. "You don't have to get up, mornings, and you don't have to go to school, and wash, and all that blame foolishness. You see a pirate don't have to do anything, Joe, when he's ashore, but a hermit he has to be praying considerable, and then he don't have any fun, anyway, all by himself that way."

"Oh, yes, that's so," said Joe, "but I hadn't thought much about it, you know. I'd a good deal rather be a pirate, now that I've tried it."

"You see," said Tom, "people don't go much on hermits, nowadays, like they used to in old times, but a pirate's always respected. And a hermit's got to sleep on the hardest place he can find, and put sackcloth and ashes on his head, and stand out in the rain, and—"

"What does he put sackcloth and ashes on his head for?" inquired Huck.

"I dunno. But they've got to do it. Hermits always do. You'd have to do that if you was a hermit."

"Derned if I would," said Huck.

"Well, what would you do?"

"I dunno. But I wouldn't do that."

"Why Huck, you'd have to. How'd you get around it?"

"Why, I just wouldn't stand it. I'd run away."

"Run away! Well, you would be a nice old slouch of a hermit. You'd be a disgrace."

The Red-Handed made no response, being better employed. He had finished gouging out a cob, and now he fitted a weed stem to it, loaded it with tobacco, and was pressing a coal to the charge and blowing a cloud of fragrant smoke—he was in the full bloom of luxurious contentment. The other pirates envied him this majestic vice, and secretly resolved to acquire it shortly. Presently Huck said:

"What does pirates have to do?"

Tom said:

"Oh, they have just a bully time—take ships and burn them, and get the money and bury it in awful places in their island where there's ghosts and things to watch it, and kill everybody in the ships—make 'em walk a plank."

"And they carry the women to the island," said Joe; "they don't kill the women."

"No," assented Tom, "they don't kill the women—they're too noble. And the women's always beautiful, too."

"And don't they wear the bulliest clothes! Oh, no! All gold and silver and di'monds," said Joe, with enthusiasm.

"Who?" said Huck.

"Why, the pirates."

Huck scanned his own clothing forlornly.

"I reckon I ain't dressed fitten for a pirate," said he, with a regretful pathos in his voice; "but I ain't got none but these."

But the other boys told him the fine clothes would come fast enough, after they should have begun their adventures. They made him understand that his poor rags would do to begin with, though it was customary for wealthy pirates to start with a proper wardrobe.

Gradually their talk died out and drowsiness began to steal upon the eyelids of the little waifs. The pipe dropped from the fingers of the Red-Handed, and he slept the sleep of the conscience-free and the weary. The Terror of the Seas and the Black Avenger of the Spanish Main had more difficulty in getting to sleep. They said their prayers inwardly, and lying down, since there was nobody there with authority to make them kneel and recite aloud; in truth, they had a mind not to say them at all, but they were afraid to proceed to such lengths as that, lest they might call down a sudden and special thunderbolt from heaven. Then at once they reached and hovered upon the imminent verge of sleep—but an intruder came, now, that would not "down." It was conscience. They began to feel a vague fear that they had been doing wrong to run away, and next they thought of the stolen meat, and then the real torture came. They tried to argue it away by reminding conscience that they had purloined sweetmeats and apples scores of times; but conscience was not to be appeased by such thin plausibilities; it seemed to them, in the end, that there was no getting around the stubborn fact that taking sweetmeats was only "hooking," while taking bacon and hams and such valuables was plain simple stealing—and there was a command against that in the Bible. So they inwardly resolved that so long as they remained in the business, their piracies should not again be sullied with the crime of stealing. Then conscience granted a truce, and these curiously inconsistent pirates fell peacefully to sleep.

CHAPTER 11

"IN WHICH A CONVERSATION TAKES PLACE WHICH SEEMS LIKELY TO COST PHILEAS FOGG DEAR" FROM *AROUND THE WORLD IN EIGHTY DAYS*

by Jules Verne

Phileas Fogg, having shut the door of his house at half-past eleven, and having put his right foot before his left five hundred and seventy-five times, and his left foot before his right five hundred and seventy-six times, reaches the Reform Club, an imposing edifice in Pall Mall, which could not have cost less than three million. He repaired at once to the dining-room, the nine windows of which open upon a tasteful garden, where the trees were already gilded with an autumn colouring; and took his place at the habitual table, the cover of which had already been laid for him. His breakfast consisted of a side-dish, a broiled fish with Reading sauce, a scarlet slice of roast beef garnished with mushrooms, a rhubarb and gooseberry tart, and a morsel of Cheshire cheese, the whole being washed down with several cups of tea, for which the Reform is famous. He rose at thirteen minutes to one, and directed his steps towards the large hall, a sumptuous apartment adorned with lavishly-framed paintings. A flunkey handed him an uncut Times, which he proceeded to cut with a skill which betrayed familiarity with this delicate operation. The perusal of this paper absorbed Phileas Fogg until a quarter before four, whilst the Standard, his next task, occupied him till the dinner hour. Dinner passed as breakfast had done, and Mr. Fogg reappeared in the reading-room and sat down to the Pall Mall at twenty minutes before six. Half an hour later several members of the Reform came in and drew up to the fireplace, where a coal fire was steadily burning. They were Mr. Fogg's usual partners at whist: Andrew Stuart, an engineer; John Sullivan and Samuel Fallentin, bankers; Thomas Flanagan, a brewer; and Gauthier Ralph, one of the Directors of the Bank of England—all rich and highly respectable personages, even in a club which comprises the princes of English trade and finance.

"Well, Ralph," said Thomas Flanagan, "what about that robbery?"

"Oh," replied Stuart, "the Bank will lose the money."

"On the contrary," broke in Ralph, "I hope we may put our hands on the robber. Skillful detectives have been sent to all the principal ports of America and the Continent, and he'll be a clever fellow if he slips through their fingers."

But have you got the robber's description?" asked Stuart.

"In the first place, he is no robber at all," returned Ralph, positively.

"What! a fellow who makes off with fifty-five thousand pounds, no robber?"

"No."

"Perhaps he's a manufacturer, then."

"*The Daily Telegraph* says that he is a gentleman."

It was Phileas Fogg, whose head now emerged from behind his newspapers, who made this remark. He bowed to his friends, and entered into the conversation. The affair which formed its subject, and which was town talk, had occurred three days before at the Bank of England. A package of banknotes, to the value of fifty-five thousand pounds, had been taken from the principal cashier's table, that functionary being at the moment engaged in registering the receipt of three shillings and sixpence. Of course, he could not have his eyes everywhere. Let it be observed that the Bank of England reposes a touching confidence in the honesty of the public. There are neither guards nor gratings to protect its treasures; gold, silver, banknotes are freely exposed, at the mercy of the first comer. A keen observer of English customs relates that, being in one of the rooms of the Bank one day, he had the curiosity to examine a gold ingot weighing some seven or eight pounds. He took it up, scrutinised it, passed it to his neighbor, he to the next man, and so on until the ingot, going from hand to hand, was transferred to the end of a dark entry; nor did it return to its place for half an hour. Meanwhile, the cashier had not so much as raised his head. But in the present instance things had not gone so smoothly. The package of notes not being found when five o'clock sounded from the ponderous clock in the "drawing office," the amount was passed to the account of profit and loss. As soon as the robbery was discovered, picked detectives hastened off to Liverpool, Glasgow, Havre, Suez, Brindisi, New York, and other ports, inspired by the proffered reward of two thousand pounds, and five per cent on the sum that might be recovered. Detectives were also charged with narrowly watching those who arrived at or left London by rail, and a judicial examination was at once entered upon.

There were real grounds for supposing, as the *Daily Telegraph* said, that the thief did not belong to a professional band. On the day of the robbery a well-dressed gentleman of polished manners, and with a well-to-do air, had been observed going to and fro in the paying room where the crime was committed. A description of him was easily procured and sent to the detectives; and some hopeful spirits, of whom Ralph was one, did not despair of his apprehension. The papers and clubs were full of the affair, and everywhere people were discussing the probabilities of a successful pursuit; and the Reform Club was especially agitated, several of its members being Bank officials.

Ralph would not concede that the work of the detectives was likely to be in vain, for he thought that the prize offered would greatly stimulate their zeal and activity. But Stuart was far from sharing this confidence; and, as they placed themselves at the whist-table, they continued to argue the matter. Stuart and Flanagan played together, while Phileas Fogg had Fallentin for his partner. As the game proceeded the conversation ceased, excepting between the rubbers, when it revived again.

"I maintain," said Stuart, "that the chances are in favour of the thief, who must be a shrewd fellow."

"Well, but where can he fly to?" asked Ralph. "No country is safe for him."

"Pshaw!"

"Where could he go, then?"

"Oh, I don't know that. The world is big enough."

"It was once," said Phileas Fogg, in a low tone. "Cut, sir," he added, handing the cards to Thomas Flanagan.

The discussion fell during the rubber, after which Stuart took up its thread.

"What do you mean 'once'? Has the world grown smaller?"

"Certainly," returned Ralph. "I agree with Mr. Fogg. The world has grown smaller, since a man can now go round it ten times more quickly than a hundred years ago. And that is why the search for this thief will be more likely to succeed."

"And also why the thief can get away more easily."

"Be so good as to play, Mr. Stuart," said Phileas Fogg.

But the incredulous Stuart was not convinced, and when the hand was finished, said eagerly: "You have a strange way, Ralph, of proving that the world has grown smaller. So, because you can go round it in three months—"

"In eighty days," interrupted Phileas Fogg.

"That is true, gentlemen," added John Sullivan. "Only eighty days, now that the section between Rothal and Allahabad, on the Great Indian Peninsula Railway, has been opened. Here is the estimate made by the *Daily Telegraph*.

- From London to Suez via Mont Cenis and Brindisi, by rail and steamboats: 7 days
- From Suez to Bombay, by steamer: 13 days
- From Bombay to Calcutta, by rail: 3 days

- From Calcutta to Hong Kong, by steamer: 13 days
- From Hong Kong to Yokohama (Japan), by steamer: 6 days
- From Yokohama to San Francisco, by steamer: 22 days
- From San Francisco to New York, by rail: 7 days
- From New York to London, by steamer and rail: 9 days
- Total: 80 days

"Yes, in eighty days!" exclaimed Stuart, who in his excitement made a false deal. "But that doesn't take into account bad weather, contrary winds, shipwrecks, railway accidents, and so on."

"All included," returned Phileas Fogg, continuing to play despite the discussion.

"But suppose the Hindoos or Indians pull up the rails," replied Stuart; "suppose they stop the trains, pillage the luggage-vans, and scalp the passengers!"

"All included," calmly retorted Fogg; adding, as he threw down the cards, "Two trumps."

Stuart, whose turn it was to deal, gathered them up, and went on: "You are right, theoretically, Mr. Fogg, but practically— "

"Practically also, Mr. Stuart."

"I'd like to see you do it in eighty days."

"It depends on you. Shall we go?"

"Heaven preserve me! But I would wager four thousand pounds that such a journey, made under these conditions, is impossible."

"Quite possible, on the contrary," returned Mr. Fogg.

"Well, make it, then!"

"The journey round the world in eighty days?"

"Yes."

"I should like nothing better."

"When?"

"At once. Only I warn you that I shall do it at your expense."

"It's absurd!" cried Stuart, who was beginning to be annoyed at the persistency of his friend. "Come, let's go on with the game."

"Deal over again, then," said Phileas Fogg. "There's a false deal."

Stuart took up the pack with a feverish hand; then suddenly put them down again.

"Well, Mr. Fogg," said he, "it shall be so: I will wager the four thousand on it."

"Calm yourself, my dear Stuart," said Fallentin. "It's only a joke."

"When I say I'll wager," returned Stuart, "I mean it." "All right," said Mr. Fogg; and, turning to the others, he continued: "I have a deposit of twenty thousand at Baring's which I will willingly risk upon it."

"Twenty thousand pounds!" cried Sullivan. "Twenty thousand pounds, which you would lose by a single accidental delay!"

"The unforeseen does not exist," quietly replied Phileas Fogg.

"But, Mr. Fogg, eighty days are only the estimate of the least possible time in which the journey can be made."

"A well-used minimum suffices for everything."

"But, in order not to exceed it, you must jump mathematically from the trains upon the steamers, and from the steamers upon the trains again."

"I will jump—mathematically."

"You are joking."

"A true Englishman doesn't joke when he is talking about so serious a thing as a wager," replied Phileas Fogg, solemnly. "I will bet twenty thousand pounds against anyone who wishes that I will make the tour of the world in eighty days or less; in nineteen hundred and twenty hours, or a hundred and fifteen thousand two hundred minutes. Do you accept?"

"We accept," replied Messrs. Stuart, Fallentin, Sullivan, Flanagan, and Ralph, after consulting each other.

"Good," said Mr. Fogg. "The train leaves for Dover at a quarter before nine. I will take it."

"This very evening?" asked Stuart.

"This very evening," returned Phileas Fogg. He took out and consulted a pocket almanac, and added, "As today is Wednesday, the 2nd of October, I shall be due in London at this very room of the Reform Club, on Saturday, the 21st of December, at a quarter before nine p.m.; or else the twenty thousand pounds, now deposited in my name at Baring's, will belong to you, in fact and in right, gentlemen. Here is a cheque for the amount."

A memorandum of the wager was at once drawn up and signed by the six parties, during which Phileas Fogg preserved a stoical composure. He certainly did not bet to win, and had only staked the twenty thousand pounds, half of his fortune, because he foresaw that he might have to expend the other half to carry out this difficult, not to say unattainable, project. As for his antagonists, they seemed much agitated;

not so much by the value of their stake, as because they had some scruples about betting under conditions so difficult to their friend.

The clock struck seven, and the party offered to suspend the game so that Mr. Fogg might make his preparations for departure.

"I am quite ready now," was his tranquil response. "Diamonds are trumps: be so good as to play, gentlemen."

CHAPTER 12

HOW THE BRIGADIER GENERAL SLEW THE FOX

by Arthur Conan Doyle

In all the great hosts of France there was only one officer toward whom the English of Wellington's Army retained a deep, steady, and unchangeable hatred.

There were plunderers among the French, and men of violence, gamblers, duellists, and roués. All these could be forgiven, for others of their kind were to be found among the ranks of the English. But one officer of Massena's force had committed a crime which was unspeakable, unheard of, abominable; only to be alluded to with curses late in the evening, when a second bottle had loosened the tongues of men. The news of it was carried back to England, and country gentlemen who knew little of the details of the war grew crimson with passion when they heard of it, and yeomen of the shires raised freckled fists to Heaven and swore. And yet who should be the doer of this dreadful deed but our friend the Brigadier, Etienne Gerard, of the Hussars of Conflans, gay-riding, plume-tossing, debonair, the darling of the ladies and of the six brigades of light cavalry.

But the strange part of it is that this gallant gentleman did this hateful thing, and made himself the most unpopular man in the Peninsula, without ever knowing that he had done a crime for which there is hardly a name amid all the resources of our language. He died of old age, and never once in that imperturbable self-confidence which adorned or disfigured his character knew that so many thousand Englishmen would gladly have hanged him with their own hands. On the contrary, he numbered this adventure among those other exploits which he has given to the world, and many a time he chuckled and hugged himself as he narrated it to the eager circle who gathered round him in that humble cafe where, between his dinner and his dominoes, he would tell, amid tears and laughter, of that inconceivable Napoleonic past when France, like and angel of wrath, rose up, splendid and terrible, before a cowering continent. Let us listen to him as he tells the story in his own way and from his own point of view.

You must know, my friends, said he, that it was toward the end of the year eighteen hundred and ten that I and Massena and the others pushed Wellington backward until we had hoped to drive him and his army into the Tagus. But when we were still twenty five miles from Lisbon we found that we were betrayed, for what had this Englishman done but build an enormous line of works and forts at a place called Torres Vedras, so that even we were unable to get through them! They lay across the whole Peninsula, and our army was so far from home that we did not dare to risk a reverse, and we had already learned at Busaco that it was no child's play to fight against these people. What could we do, then, but sit down in front of these lines and blockade them to the best of our power? There we remained for six months, amid such anxieties that Massena said afterward that he had not one hair which was not white upon his body.

For my own part, I did not worry much about our situation, but I looked after our horses, who were in much need of rest and green fodder. For the rest, we drank the wine of the country and passed the time as best we might. There was lady at Santarem—but my lips are sealed. It is the part of a gallant man to say nothing, though he may indicate that he could say a great deal.

One day Massena sent for me, and I found him in his tent with a great plan pinned upon the table. He looked at me in silence with that single piercing eye of his, and I felt by his expression that the matter was serious. He was nervous and ill at ease, but my bearing seemed to reassure him. It is good to be in contact with brave men.

"Colonel Etienne Gerard," said he, "I have always heard that you are a very gallant and enterprising officer."

It was not for me to confirm such a report, and yet it would be folly to deny it, so I clinked my spurs together and saluted.

"You are also an excellent rider."

I admitted it.

"And the best swordsman in the six brigades of light cavalry."

Massena was famous for the accuracy of his information.

"Now," said he, "if you will look at this plan you will have no difficulty in understanding what it is that I wish you to do. These are the lines of Torres Vedras. You will perceive that they cover a vast space, and you will realise that the English can only hold a position here and there. Once through the lines you have twenty-five miles of open country which lie between them and Lisbon. It is very important to me to learn how Wellington's troops are distributed throughout that space, and it is my wish that you should go and ascertain."

His words turned me cold.

"Sir," said I, "it is impossible that a colonel of light cavalry should condescend to act as a spy."

He laughed and clapped me on the shoulder.

"You would not be a Hussar if you were not a hot-head," said he. "If you will listen you will understand that I have not asked you to act as a spy. What do you think of that horse?"

He had conducted me to the opening of his tent, and there was a chasseur who led up and down a most admirable creature. He was a dapple grey, not very tall, a little over fifteen hands perhaps, but with the short head and splendid arch of the neck which comes with the Arab blood. His shoulders and haunches were so muscular, and yet his legs so fine, that it thrilled me with joy just to gaze upon him. A fine horse or a beautiful woman—I cannot look at them unmoved, even now when seventy winters have chilled my blood. You can think how it was in the year' 10.

"This," said Massena, "is Voltigeur, the swiftest horse in our army. What I desire is that you should start tonight, ride round the lines upon the flank, make your way across the enemy's rear, and return upon the other flank, bringing me news of his disposition. You will wear a uniform, and will, therefore, if captured, be safe from the death of a spy. It is probable that you will get through the lines unchallenged, for the posts are very scattered. Once through, in daylight you can outride anything which you meet, and if you keep off the roads you may escape entirely unnoticed. If you have not reported yourself by tomorrow night, I will understand that you are taken, and I will offer them Colonel Petrie in exchange."

Ah, how my heart swelled with pride and joy as I sprang into the saddle and galloped this grand horse up and down to show the Marshal the mastery which I had of him! He was magnificent—we were both magnificent, for Massena clapped his hands and cried out in his delight.

It was not I, but he, who said that a gallant beast deserves a gallant rider. Then, when for the third time, with my panache flying and my dolman streaming behind me, I thundered past him, I saw upon his hard old face that he had no longer any doubt that he had chosen the man for his purpose. I drew my sabre, raised the hilt to my lips in salute, and galloped on to my own quarters.

Already the news had spread that I had been chosen for a mission, and my little rascals came swarming out of their tents to cheer me. Ah! It brings the tears to my old eyes when I think how proud they were of their Colonel.

And I was proud of them also. They deserved a dashing leader.

The night promised to be a stormy one, which was very much to my liking. It was my desire to keep my departure most secret, for it was evident that if the English heard that I had been detached from the army they would naturally conclude that something important was about to happen. My horse was taken, therefore, beyond the picket line, as if for watering, and I followed and mounted him there. I had a map, a compass, and a paper of instructions from the Marshal, and with these in the bosom of my tunic and my sabre at my side I set out upon my adventure.

A thin rain was falling and there was no moon, so you may imagine that it was not very cheerful. But my heart was light at the thought of the honour which had been done me and the glory which awaited me. This exploit should be one more in that brilliant series which was to change my sabre into a baton. Ah, how we dreamed, we foolish fellows, young, and drunk with success! Could I have foreseen that night as I rode, the chosen man of sixty thousand, that I should spend my life planting cabbages on a hundred francs a month! Oh, my youth, my hopes, my comrades! But the wheel turns and never stops. Forgive me, my friends, for an old man has his weakness.

My route, then, lay across the face of the high ground of Torres Vedras, then over a streamlet, past a farmhouse which had been burned down and was now only a landmark, then through a forest of young cork oaks, so to the monastery of San Antonio, which marked the left of the English position. Here I turned south and rode quietly over the downs, for it was at this point that Massena thought that it would be most easy for me to find my way unobserved through the position. I went very slowly, for it was so dark that I could not see my hand in front of me. In such cases I leave my bridle loose and let my horse pick its own way. Voltigeur went confidently forward, and I was very content to sit upon his back and to peer about me, avoiding every light.

For three hours we advanced in this cautious way, until it seemed to me that I must have left all danger behind me. I then pushed on more briskly, for I wished to be in the rear of the whole army by daybreak. There are many vineyards in these parts which in winter become open plains, and a horseman finds few difficulties in his way.

But Massena had underrated the cunning of these English, for it appears that there was not one line of defence but three, and it was the third, which was the most formidable, through which I was at that instant passing. As I rode, elated at my own success, a lantern flashed suddenly before me, and I saw the glint of polished gun barrels and the gleam of a red coat.

"Who goes there?" cried a voice—such a voice! I swerved to the right and rode like a madman, but a dozen squirts of fire came out of the darkness, and the bullets whizzed all round my ears. That was no new sound to me, my friends, though I will not talk like a foolish conscript and say that I have ever liked it. But at least it had never kept me from thinking clearly, and so I knew that there was nothing for it but to gallop hard and try my luck elsewhere. I rode round the English picket, and then, as I heard more of them, I concluded rightly that I had at last come through their defences.

For five miles I rode south, striking a tinder from time to time to look at my pocket compass. And then in an instant—I feel the pang once more as my memory brings back the moment—my horse, without a sob or staggers fell stone-dead beneath me!

I had never known it, but one of the bullets from that infernal picket had passed through his body. The gallant creature had never winced nor weakened, but had gone while life was in him. One instant I was secure on the swiftest, most graceful horse in Massena's army. The next he lay upon his side, worth only the price of his hide, and I stood there that most helpless, most ungainly of creatures, a dismounted Hussar. What could I do with my boots, my spurs, my trailing sabre? I was far inside the enemy's lines. How could I hope to get back again?

I am not ashamed to say that I, Etienne Gerard, sat upon my dead horse and sank my face in my hands in my despair.

Already the first streaks were whitening the east.

In half an hour it would be light. That I should have won my way past every obstacle and then at this last instant be left at the mercy of my enemies, my mission ruined, and myself a prisoner—was it not enough to break a soldier's heart?

But courage, my friends! We have these moments of weakness,

the bravest of us; but I have a spirit like a slip of steel, for the more you bend it the higher it springs.

One spasm of despair, and then a brain of ice and a heart of fire. All was not yet lost. I who had come through so many hazards would come through this one also. I rose from my horse and considered what had best be done.

And first of all it was certain that I could not get back. Long before I could pass the lines it would be broad daylight. I must hide myself for the day, and then devote the next night to my escape. I took the saddle, holsters, and bridle from poor Voltigeur, and I concealed them among some bushes, so that no one finding him could know that he was a French horse. Then, leaving him lying there, I wandered on in search of some place where I might be safe for the day. In every direction I could see camp fires upon the sides of the hills, and already figures had begun to move around them. I must hide quickly, or I was lost.

But where was I to hide? It was a vineyard in which I found myself, the poles of the vines still standing, but the plants gone. There was no cover there. Besides, I should want some food and water before another night had come. I hurried wildly onward through the waning darkness, trusting that chance would be my friend.

And I was not disappointed. Chance is a woman, my friends, and she has her eye always upon a gallant Hussar.

Well, then, as I stumbled through the vineyard, something loomed in front of me, and I came upon a great square house with another long, low building upon one side of it. Three roads met there, and it was easy to see that this was the posada, or wine shop.

There was no light in the windows, and everything was dark and silent, but of course, I knew that such comfortable quarters were certainly occupied, and probably by someone of importance. I have learned, however, that the nearer the danger may really be the safer place, and so I was by no means inclined to trust myself away from this shelter. The low building was evidently the stable, and into this I crept, for the door was unlatched.

The place was full of bullocks and sheep, gathered there, no doubt, to be out of the clutches of marauders.

A ladder led to a loft, and up this I climbed and concealed myself very snugly among some bales of hay upon the top. This loft had a small open window, and I was able to look down upon the front of the inn and also upon the road. There I crouched and waited to see what would happen.

It was soon evident that I had not been mistaken when I had thought that this might be the quarters of some person of importance. Shortly after daybreak an English light dragoon arrived with a despatch, and from them onward the place was in a turmoil, officers continually riding up and away. Always the same name was upon their lips: "Sir Stapleton—Sir Stapleton."

It was hard for me to lie there with a dry moustache and watch the great flagons which were brought out by the landlord to these English officers. But it amused me to look at their fresh-coloured, clean-shaven, careless faces, and to wonder what they would think if they knew that so celebrated a person was lying so near to them. And then, as I lay and watched, I saw a sight which filled me with surprise.

It is incredible the insolence of these English! What do you suppose Milord Wellington had done when he found that Massena had blockaded him and that he could not move his army? I might give you many guesses. You might say that he had raged, that he had despaired, that he had brought his troops together and spoken to them about glory and the fatherland before leading them to one last battle. No, Milord did none of these things. But he sent a fleet ship to England to bring him a number of fox-dogs; and he with his officers settled himself down to chase the fox. It is true what I tell you. Behind the lines of Torres Vedras these mad Englishmen made the fox chase three days in the week.

We had heard of it in the camp, and now I was myself to see that it was true.

For, along the road which I have described, there came these very dogs, thirty or forty of them, white and brown, each with its tail at the same angle, like the bayonets of the Old Guard. My faith, but it was a pretty sight! And behind and amidst them there rode three men with peaked caps and red coats, whom I understood to be the hunters. After them came many horsemen with uniforms of various kinds, stringing along the roads in twos and threes, talking together and laughing.

They did not seem to be going above a trot, and it appeared to me that it must indeed be a slow fox which they hoped to catch. However, it was their affair, not mine, and soon they had all passed my window and were out of sight. I waited and I watched, ready for any chance which might offer.

Presently an officer, in a blue uniform not unlike that of our flying artillery, came cantering down the road—an elderly, stout man he was, with grey side-whiskers. He stopped and began to talk with an orderly officer of dragoons, who waited outside the inn, and it was then that I learned the advantage of the English which had been taught me. I could hear and understand all that was said.

"Where is the meet?" said the officer, and I thought that he was hungering for his bifstek. But the other answered him that it was near Altara, so I saw that it was a place of which he spoke.

"You are late, Sir George," said the orderly.

"Yes, I had a court-martial. Has Sir Stapleton Cotton gone?"

At this moment a window opened, and a handsome young man in a very splendid uniform looked out of it.

"Halloa, Murray!" said he. "These cursed papers keep me, but I will be at your heels."

"Very good, Cotton. I am late already, so I will ride on."

"You might order my groom to bring round my horse," said the young General at the window to the orderly below, while the other went on down the road.

The orderly rode away to some outlying stable, and then in a few minutes there came a smart English groom with a cockade in his hat, leading by the bridle a horse—and, oh, my friends, you have never known the perfection to which a horse can attain until you have seen a first-class English hunter. He was superb: tall, broad, strong, and yet as graceful and agile as a deer. Coal black he was in colour, and his neck, and his shoulder, and his quarters, and his fetlocks—how can I describe him all to you? The sun shone upon him as on polished ebony, and he raised his hoofs in a little playful dance so lightly and prettily, while he tossed his mane and whinnied with impatience. Never have I seen such a mixture of strength and beauty and grace. I had often wondered how the English Hussars had managed to ride over the chasseurs of the Guards in the affair at Astorga, but I wondered no longer when I saw the English horses.

There was a ring for fastening bridles at the door of the inn, and the groom tied the horse there while he entered the house. In an instant I had seen the chance which Fate had brought to me. Were I in that saddle I should be better off than when I started. Even Voltigeur could not compare with this magnificent creature. To think is to act with me. In one instant I was down the ladder and at the door of the stable. The next I was out and the bridle was in my hand. I bounded into the saddle.

Somebody, the master or the man, shouted wildly behind me. What cared I for his shouts! I touched the horse with my spurs and he bounded forward with such a spring that only a rider like myself could have sat him. I gave him his head and let him go—it did not matter to me where, so long as we left this inn far behind us. He thundered away across the vineyards, and in a very few minutes I had placed miles between myself and my pursuers. They could no longer tell in that wild country in which direction I had gone. I knew that I was safe, and so, riding to the top of a small hill, I drew my pencil and notebook from my pocket and proceeded to make plans of those camps which I could see and to draw the outline of the country.

Foxhunting: Encouraging Hound by John Frederick Herring, Sr., 1839. Courtesy of the Yale Center of British Art.

He was a dear creature upon whom I sat, but it was not easy to draw upon his back, for every now and then his two ears would cock, and he would start and quiver with impatience. At first I could not understand this trick of his, but soon I observed that he only did it when a peculiar noise—*yoy, yoy, yoy*—came from somewhere among the oak woods beneath us. And then suddenly this strange cry changed into a most terrible SCREAMING, with the frantic blowing of a horn. Instantly he went mad—this horse. His eyes blazed. His mane bristled. He bounded from the earth and bounded again, twisting and turning in a frenzy. My pencil flew one way and my notebook another. And then, as I looked down into the valley, an extraordinary sight met my eyes.

The hunt was streaming down it. The fox I could not see, but the dogs were in full cry, their noses down, their tails up, so close together that they might have been one great yellow and white moving carpet. And behind them rode the horsemen—my faith, what a sight! Consider every type which a great army could show. Some in hunting dress, but the most in uniforms: blue dragoons, red dragoons, redtrousered hussars, green riflemen, artillerymen, goldslashed lancers, and most of all red, red, red, for the infantry officers ride as hard as the cavalry.

Such a crowd, some well-mounted, some ill, but all flying along as best they might, the subaltern as good as the general, jostling and pushing, spurring and driving, with every thought thrown to the winds save that they should have the blood of this absurd fox! Truly, they are an extraordinary people, the English!

But I had little time to watch the hunt or to marvel at these islanders, for of all these mad creatures the very horse upon which I sat was the maddest. You understand that he was himself a hunter, and that the crying of these dogs was to him what the call of a cavalry trumpet in the street yonder would be to me. It thrilled him. It drove him wild. Again and again he bounded into the air, and then, seizing the bit between his teeth, he plunged down the slope and galloped after the dogs.

I swore, and tugged, and pulled, but I was powerless.

This English General rode his horse with a snaffle only, and the beast had a mouth of iron. It was useless to pull him back. One might as well try to keep a grenadier from a wine bottle. I gave it up in despair, and, settling down in the saddle, I prepared for the worst which could befall.

What a creature he was! Never have I felt such a horse between my knees. His great haunches gathered under him with every stride, and he shot forward ever faster and faster, stretched like a greyhound, while the wind beat in my face and whistled past my ears. I was wearing our undress jacket, a uniform simple and dark in itself—though some figures give distinction to any uniform—and I had taken the precaution to remove the long panache from my busby. The result was that, amidst the mixture of costumes in the hunt, there was no reason why mine should attract attention, or why these men, whose thoughts were all with the chase, should give any heed to me. The idea that a French officer might be riding with them was too absurd to enter their minds. I laughed as I rode, for indeed, amid all the danger, there was something of comic in the situation.

I have said that the hunters were very unequally mounted, and so at the end of a few miles, instead of being one body of men, like a charging regiment, they were scattered over a considerable space, the better riders well up to the dogs and the others trailing away behind.

Now, I was as good a rider as any, and my horse was the best of them all, and so you can imagine that it was not long before he carried me to the front. And when I saw the dogs streaming over the open, and the red-coated huntsman behind them, and only seven or eight horsemen between us, then it was that the strangest thing of all happened, for I, too, went mad—I, Etienne Gerard!

In a moment it came upon me, this spirit of sport, this desire to excel, this hatred of the fox. Accursed animal, should he then defy us? Vile robber, his hour was come!

Ah, it is a great feeling, this feeling of sport, my friends, this desire to trample the fox under the hoofs of your horse. I have made the fox chase with the English. I have also, as I may tell you some day, fought the box-fight with the Bustler, of Bristol.

And I say to you that this sport is a wonderful thing—full of interest as well as madness.

The farther we went the faster galloped my horse, and soon there were but three men as near the dogs as I was.

All thought of fear of discovery had vanished. My brain throbbed, my blood ran hot—only one thing upon earth seemed worth living for, and that was to overtake this infernal fox. I passed one of the horsemen—a Hussar like myself. There were only two in front of me now: the one in a black coat, the other the blue artilleryman whom I had seen at the inn. His grey whiskers streamed in the wind, but he rode magnificently. For a mile or more we kept in this order, and then, as we galloped up a steep slope, my lighter weight brought me to the front.

I passed them, both, and when I reached the crown I was riding level with the little, hard-faced English huntsman.

In front of us were the dogs, and then, a hundred paces beyond them, was a brown wisp of a thing, the fox itself, stretched to the uttermost. The sight of him fired my blood. "Aha, we have you then, assassin!" I cried, and shouted my encouragement to the huntsman. I waved my hand to show him that there was one upon whom he could rely.

And now there were only the dogs between me and my prey. These dogs, whose duty it is to point out the game, were now rather a hindrance than a help to us, for it was hard to know how to pass them. The huntsman felt the difficulty as much as I, for he rode behind them, and could make no progress toward the fox. He was a swift rider, but wanting in enterprise. For my part, I felt that it would be unworthy of the Hussars of Conflans if I could not overcome such a difficulty as this.

Was Etienne Gerard to be stopped by a herd of fox-dogs?

It was absurd. I gave a shout and spurred my horse.

"Hold hard, sir! Hold hard!" cried the huntsman.

He was uneasy for me, this good old man, but I reassured him by a wave and a smile. The dogs opened in front of me. One or two may have been hurt, but what would you have? The egg must be broken for the omelette. I could hear the huntsman shouting his congratulations behind me. One more effort, and the dogs were all behind me. Only the fox was in front.

Ah, the joy and pride of that moment! To know that I had beaten the English at their own sport. Here were three hundred, all thirsting for the life of this animal, and yet it was I who was about to take it. I thought of my comrades of the light cavalry brigade, of my mother, of the Emperor, of France. I had brought honour to each and all. Every instant brought me nearer to the fox. The moment for action had arrived, so I unsheathed my sabre. I waved it in the air, and the brave English all shouted behind me.

Only then did I understand how difficult is this fox chase, for one may cut again and again at the creature and never strike him once. He is small, and turns quickly from a blow. At every cut I heard those shouts of encouragement from behind me, and they spurred me to yet another effort. And then at last the supreme moment of my triumph arrived. In the very act of turning I caught him fair with such another back-handed cut as that with which I killed the aide-de-camp of the Emperor of Russia. He flew into two pieces, his head one way and his tail another. I looked back and waved the blood-stained sabre in the air. For the moment I was exalted—superb!

Ah! How I should have loved to have waited to have received the congratulations of these generous enemies.

There were fifty of them in sight, and not one who was not waving his hand and shouting. They are not really such a phlegmatic race, the English. A gallant deed in war or sport will always warm their hearts. As to the old huntsman, he was the nearest to me, and I could see with my own eyes how overcome he was by what he had seen. He was like a man paralysed, his mouth open, his hand, with outspread fingers, raised in the air. For a moment my inclination was to return and to embrace him.

But already the call of duty was sounding in my ears, and these English, in spite of all the fraternity which exits among sportsmen, would certainly have made me prisoner. There was no hope for my mission now, and I had done all that I could do. I could see the lines of Massena's camp no very great distance off, for by a lucky chance, the chase had taken us in that direction.

I turned from the dead fox, saluted with my sabre, and galloped away.

But they would not leave me so easily, these gallant huntsmen. I was the fox now, and the chase swept bravely over the plain. It was only at the moment when I started for the camp that they could have known that I was a Frenchman, and now the whole swarm of them were at my heels. We were within gunshot of our pickets before they would halt, and then they stood in knots and would not go away, but shouted and waved their hands at me. No, I will not think that it was in enmity. Rather would I fancy that a glow of admiration filled their breasts, and that their one desire was to embrace the stranger who had carried himself so gallantly and well.